INFINITE ABELIAN GROUPS

REVISED EDITION

IRVING KAPLANSKY

DOVER PUBLICATIONS, INC.
MINEOLA, NEW YORK

Bibliographical Note

This Dover edition, first published in 2018, is an unabridged republication of the second edition of the work, originally published by The University of Michigan Press, Ann Arbor, in 1969 [first edition: 1954]. The present edition is reprinted by special arrangement with The University of Michigan Press.

Library of Congress Cataloging-in-Publication Data

Names: Kaplansky, Irving, 1917–2006, author.
Title: Infinite Abelian groups / Irving Kaplansky.
Description: Revised edition, Dover edition. | Mineola, New York : Dover Publications, 2018. | Revised edition originally published: Ann Arbor : University of Michigan Press, 1969. | Includes bibliographical references and index.
Identifiers: LCCN 2018011854| ISBN 9780486828503 | ISBN 0486828506
Subjects: LCSH: Abelian groups.
Classification: LCC QA171 .K35 2018 | DDC 512/.25—dc23
LC record available at https://lccn.loc.gov/2018011854

Manufactured in the United States by LSC Communications
82850601 2018
www.doverpublications.com

PREFACE TO THE REVISED EDITION

In the second edition a number of misprints and errors have been corrected and Sections 6, 16, 17 (test problems, complete modules, algebraic compactness) have been extensively revised.

The new bibliography is limited to items to which there is an actual reference. The bibliography in the first edition had 145 entries and was intended to be complete up to about 1952. It would take over 400 additional entries to bring it up to 1968. It is doubtful whether the space occupied by such a large bibliography would be well spent.

The guide to the literature is omitted, but relevant parts have been incorporated into the new section of notes. For some things, e.g. my views on duality, the reader should consult the first edition.

I urge the reader to have Fuchs's definitive treatise at hand. My feeling is that there is nevertheless still room for a slim volume, not so imposing, gentle, and talkative (at least in the beginning).

I take the opportunity to disagree mildly with Professor Fuchs about the role of modules. In the applications of algebra (notably to topology), very general rings and modules over them are increasingly important. I stand by the compromise in *Infinite Abelian Groups:* start with familiar plain old abelian groups and switch completely to modules over principal ideal rings at about the half-way point.

In the appended notes there is indeed a good deal of exploration of modules, combined with remarks appropriate to a second look at the subject. These require from the reader some familiarity with the rudiments of commutative ring theory and homological algebra.

I am indebted to Peter Crawley, Alfred Hales, Charles Megibben, and Joseph Rotman for spirited comments on a draft of the second edition. Professor Fuchs kindly took time out from the preparation of his own second edition to send me valuable suggestions. In addition to the Office of Naval Research, whose aid was acknowledged in the first edition, I am happy to thank the Army Research Office, the Air Force Office of Scientific Research, and the National Science Foundation for their support over the years. Thanks also to Joyce Bolden for a splendid job of typing.

Chicago, Ill. May 1968

CONTENTS

INFINITE ABELIAN GROUPS

1. INTRODUCTION

In the early days of group theory attention was confined almost entirely to finite groups. But recently, and above all in the last two decades, the infinite group has come into its own. The results obtained on infinite *abelian* groups have been particularly penetrating. This monograph has been written with two objectives in mind: first, to make the theory of infinite abelian groups available in a convenient form to the mathematical public; second, to help students acquire some of the techniques used in modern infinite algebra.

For this second purpose infinite abelian groups serve admirably. No extensive background is required for their study, the rudiments of group theory being sufficient. There is a good variety in the transfinite tools employed, with Zorn's lemma being applied in several different ways. The traditional style of transfinite induction is not completely ignored either, for there is a theorem whose very formulation uses transfinite ordinals. The peculiar role sometimes played by a countability hypothesis makes a challenging appearance.

It is furthermore helpful that finite abelian groups are completely known. In other subjects, such as rings or nonabelian groups, there are distracting difficulties which occur even in the finite case. Here, however, our attention is concentrated on the problems arising from the fact that the groups may be infinite.

With a student audience in mind, I have given details and included remarks that would ordinarily be suppressed in print. However, as the discussion proceeds it becomes somewhat more concise. A serious effort has been made to furnish, in brief space, a reasonably complete account of the subject. In order to do this, I have relegated many results of some interest to the role of exercises, and a large part of the literature is merely surveyed in the guide to it provided in § 20.

This material is adapted from a course which I gave at the University of Chicago in the fall of 1950. I should like to record my indebtedness to the many able members of that class, particularly to George Backus, Arlen Brown, and Roger Farrell. Thanks are expressed to Isidore Fleischer for the ideas in §16 (the torsion-free case of Theorem 22 was discovered by him and appears in his doctoral dissertation); to Robert Heyneman and George Kolettis, who read a preliminary version of this work and made many valuable suggestions; to Tulane University and the University of Michigan, where I had the opportunity to lecture on abelian groups; and to the Office of Naval Research.

A special acknowledgment goes to Professor Reinhold Baer. It was from his papers that I learned much of the theory of abelian groups. Furthermore, when this monograph was nearly complete, I had the privilege of reading an unpublished manuscript (of book length) on abelian groups which he prepared in the late 1940's.

2. EXAMPLES OF ABELIAN GROUPS

Before beginning to develop the theory, it is desirable to have at hand a small collection of examples of abelian groups.

To avoid endless repetition, let it be agreed that "group" will always mean "abelian group."

(*a*) *Cyclic groups*. A group G is cyclic if it can be generated by a single element. If that element has infinite order, G is isomorphic to the additive group of integers, and is called an infinite cyclic group; if it has finite order n, G is cyclic of order n and is isomorphic to the additive group of integers mod n. We shall use the notation Z and Z_n respectively for these two groups.

(*b*) *External direct sums*. Let $\{G_i\}$ be any set of groups, where the subscript i runs over an index set I, which may be finite or infinite. We define the *direct sum* of the groups G_i. We take "vectors" $\{a_i\}$; that is, arrays indexed by $i \in I$ with a_i in G_i. Moreover, we impose the restriction that all but a finite number of the a_i's are to be 0 (we are writing 0 indifferently for the identity element of any G_i). Addition of vectors is defined by adding components. This gives an abelian group, called the direct sum of $\{G_i\}$.

If there is any danger of ambiguity, the object just defined may be referred to as the "weak," or "discrete," direct sum, as opposed to the "complete" direct sum, where the vectors are unrestricted. In pure algebra it is the weak direct sum which arises most naturally; the complete direct sum is, indeed, mostly useful as a source of counterexamples (see Theorem 21 and exercise 33).

(*c*) *Union and intersection*. If S and T are subgroups of a group, we write $S \cap T$ for their intersection, that is, the set of elements lying in both. More generally, if $\{S_i\}$ is a set of subgroups of G, we write $\cap S_i$ for the intersection. Note that we are talking about the set-theoretic intersection and that it is always a subgroup.

As regards the union of subgroups, the situation is different. Consider first two subgroups, S and T. The set-theoretic union, which we might write $S \cup T$, is not generally a subgroup (in fact, $S \cup T$ is a subgroup if and only if one of the two subgroups S and T contains the other). What we wish instead is the smallest subgroup containing S and T, and this is provided by $S + T$, the set of all elements $s + t$, where s and t range over S and T.

Again, let $\{S_i\}$ be any set of subgroups of G. Their union, written ΣS_i, is the smallest subgroup containing them; it may be explicitly described as the set of all *finite* sums of elements extracted from the various subgroups S_i.

(*d*) *Internal direct sums.* In dealing with direct sums we are most often confronted with the problem of showing that a group is isomorphic to the direct sum of certain of its subgroups. Suppose first that the group G has subgroups S and T satisfying $S \cap T = 0$, $S + T = G$. Then it is easy to see that G is isomorphic to the direct sum of S and T, where we are referring to the external direct sum discussed above in (*b*). One may speak of G as being the internal direct sum of S and T, but generally one simply calls G the direct sum of S and T, and writes $G = S \oplus T$.

Consider now any (finite or infinite) set of subgroups $\{S_i\}$. In order to verify that G is the direct sum of these subgroups, the most convenient procedure is generally as follows: Show that $G = \Sigma S_i$, that is, that every element of G can be written as a finite sum of elements from the subgroups S_i; then show that the representation is unique. This uniqueness is equivalent to the statement that each S_i is disjoint from the union of the remaining ones.

In general, if the union ΣS_i of subgroups is their direct sum, we shall call the subgroups S_i *independent.*

A concept of independence for elements will also be useful: We shall say that the elements x_i are *independent* if the cyclic subgroups they generate are independent in the sense just defined, and we write $\Sigma(x_i)$ for the subgroup generated by all the elements.

We should notice the analogy between this concept and linear independence in a vector space. In fact, the elements x_i are independent if and only if the following is true: If a finite sum

$$\Sigma n_i x_i = 0 \qquad (n_i \text{ integers}),$$

then each $n_i x_i = 0$.

(*e*) *Rational numbers.* The most general group so far in our possession is a direct sum of cyclic groups. A classical theorem asserts that this covers all finitely generated groups, and in particular all finite groups. That is to say, any finitely generated group is a direct sum of (a finite number of) cyclic groups.

One might for a moment think that perhaps any abelian group is a direct sum of cyclic groups, the number of summands now being allowed to be infinite, of course. This conjecture is defeated by a very familiar group: the additive group R of rational numbers. That R is not a direct sum of cyclic groups may be seen, for example, from the fact that for any $x \in R$ and any integer n there exists an element $y \in R$ with $ny = x$; this property manifestly cannot hold in a direct sum of cyclic groups. (The property in question is called *divisibility*, and will be studied in §5.)

(*f*) *Rationals mod one.* In the additive group R of rational numbers, there is the subgroup Z of integers. The quotient group R/Z is known as the *rationals* mod one. We note that in R/Z every element has finite order. We argue, just as above, that R/Z is not a direct sum of cyclic groups.

(*g*) *The group* $Z(p^\infty)$. There is an important modification of the two preceding examples. Let p be a fixed prime, and let P denote the additive group of those rational numbers whose denominators are powers of p. The quotient group P/Z will play a dominant role in the ensuing discussion, and we use for it the notation $Z(p^\infty)$.

Let us pause to take a close look at $Z(p^\infty)$. For simplicity we take $p = 2$. We can write the elements of $Z(2^\infty)$ as 0, 1/2, 1/4, 3/4, 1/8, etc., but it is to be understood that addition takes place mod one. Thus

$$1/2 + 1/2 = 0, \quad 1/2 + 3/4 = 1/4, \quad 3/4 + 5/8 = 3/8, \quad \text{etc.}$$

What are the subgroups of $Z(2^\infty)$? There is a subgroup of order 2 consisting of 0 and 1/2; one of order 4 consisting of 0, 1/4, 1/2, 3/4; and in general a cyclic subgroup (say H_n) of order 2^n generated by $1/2^n$. It is not difficult to see that these are in fact the only subgroups. Thus the array of subgroups can be pictured as follows:

$$0 \subset H_1 \subset H_2 \subset \cdots \subset H_n \subset \cdots \subset Z(2^\infty).$$

It is noteworthy that every subgroup of $Z(2^\infty)$ is finite, except for $Z(2^\infty)$ itself. The subgroups form an ascending chain which never terminates. On the contrary, one sees that every *descending* chain of subgroups must be finite. Thus $Z(2^\infty)$ has the so-called "descending-chain condition" but not the "ascending-chain condition."

In conclusion, we give another realization of $Z(p^\infty)$. Consider the set of all p^n-th roots of unity, where p is a fixed prime and $n = 0, 1, 2, \cdots$. These numbers form a group under multiplication, and the group is isomorphic to $Z(p^\infty)$.

This completes our discussion of examples. It will appear that these groups are the fundamental building blocks for some fairly wide classes of infinite abelian groups.

3. TORSION GROUPS

If an abelian group has all its elements of finite order, we shall call it a *torsion* group. (This designation does not convey much algebraically, but it has a suggestive topological background and the merit of brevity.) The other extreme case is that where all the elements (except 0 of course) have infinite order; we then call the group *torsion-free*.

Now let G be an arbitrary abelian group, and T the set of all elements in G having finite order. We leave to the reader the verification of the following two remarks: (*a*) T is a subgroup, (*b*) G/T is torsion-free. We shall call T the torsion subgroup of G.

The study of abelian groups is now seen to split into three parts: (*a*) the classification of torsion groups, (*b*) the classification of torsion-free groups, (*c*) the study of the way the two are put together to form an arbitrary group. Progress has been most notable on the first of these problems, and consequently we shall be chiefly concerned with torsion groups.

Next we define a group (necessarily a torsion group) to be *primary* if, for a certain prime p, every element has order a power of p. The study of torsion groups is reduced to that of primary groups by the following theorem:

Theorem 1. Any torsion group is a direct sum of primary groups.

Proof. Let G be the group, and for every prime p define G_p to be the subset consisting of elements with order a power of p. It is clear that G_p is a subgroup, and that it is primary. We shall now prove that G is isomorphic to the direct sum of the subgroups G_p.

(*a*) We have first to show that G is the union of the subgroups G_p. Take any x in G, say of order n. Then factor n into prime powers: $n = p_1^{r_1} \cdots p_k^{r_k}$, and write $n_i = n/p_i^{r_i}$ $(i = 1, \cdots, k)$. Thus $n_1, \cdots, n_k$ have greatest common divisor 1, and so there exist integers $a_1, \cdots, a_k$ with $a_1 n_1 + \cdots + a_k n_k = 1$. Then

$$x = a_1 n_1 x + \cdots + a_k n_k x. \tag{1}$$

Now $n_i x$ has precisely order $p_i^{r_i}$, and so it is in G_{p_i}. Thus equation (1) is the desired expression of x as a sum of elements in the G_p's.

(*b*) We have further to prove the uniqueness of the expression just found. Suppose

$$\begin{aligned} x &= y_1 + \cdots + y_k \\ &= z_1 + \cdots + z_k \end{aligned}$$

where y_i, z_i lie in the same G_{p_i}. Consider the equation

$$y_1 - z_1 = (z_2 + \cdots + z_k) - (y_2 + \cdots + y_k). \tag{2}$$

We know that $y_1 - z_1$ has order a power of p_1. On the other hand, the right side of (2) is an element whose order is a product of powers of $p_2, \cdots, p_k$. This is possible only if $y_1 - z_1 = 0$. Similarly each $y_i = z_i$. This completes the proof of Theorem 1.

As a general principle, every decomposition theorem should be accompanied by a uniqueness investigation. Such an investigation is particularly easy for the decomposition given by Theorem 1. In fact, there is only one way to express a torsion group as a direct sum of primary subgroups, one for each prime p; for the subgroup attached to p must necessarily consist of all elements whose order is a power of p. In other words, the decomposition is unique not just up to isomorphism; the summands are unique subgroups.

The simplicity of the proof of Theorem 1 is a natural counterpart to this strong uniqueness; for if a decomposition is unique there ought to be a simple natural way to effect it. It is instructive to compare this situation with later ones. For example, under suitable hypotheses of various kinds we shall prove that a primary group is a direct sum of cyclic groups; this decomposition is unique, but only up to isomorphism. The difficulties encountered in the proof are a natural reflection of the large number of arbitrary choices that have to be made in carrying out the decomposition.

We shall conclude this section by giving two illustrations of Theorem 1:

(*a*) Consider the cyclic group Z(n), where $n = p_1^{r_1} \cdots p_k^{r_k}$. Then $Z(n) = Z(p_1^{r_1}) \oplus \cdots \oplus Z(p_k^{r_k})$. (Indeed, this is the fact which really underlies the proof of Theorem 1.)

(*b*) Let G be the additive group of rationals mod one (§ 2). This is a torsion group, and it can be seen that its primary component for the prime p is precisely the group $Z(p^\infty)$ of § 2. Thus G is a direct sum of all the groups $Z(p^\infty)$.

4. ZORN'S LEMMA

Nearly every proof to follow will depend on the use of a transfinite induction. Such an induction is generally best accomplished by the use of Zorn's lemma, which is to be regarded as an axiom like other axioms needed to set up the foundations of mathematics.

We shall make use of a version of Zorn's lemma which refers to the concept of a partially ordered set. A partially ordered set is a set with a binary relation $\geqq$ which satisfies

(*a*) $x \geqq x$ (reflexivity),
(*b*) $x \geqq y$, $y \geqq x$ imply $x = y$ (antisymmetry),
(*c*) $x \geqq y$, $y \geqq z$ imply $x \geqq z$ (transitivity).

Let S be a partially ordered set and T a subset. The element x is said to be the least upper bound of T if $x \geqq y$ for every y in T and if $z \geqq y$ for every y in T implies $z \geqq x$. (The element x itself may or may not be in T.) A least upper bound need not exist, but if it does, it is unique.

An element x of a partially ordered set S is said to be *maximal* if S contains no larger element. It is to be observed that S may contain many maximal elements.

A partially ordered set is a *chain* (also called a simply ordered set or a linearly ordered set) if every two elements are comparable; that is, either $x \geqq y$ or $y \geqq x$.

We now state Zorn's lemma:

Zorn's lemma. Let S *be a partially ordered set in which every chain has a least upper bound. Then* S *has a maximal element.*

This brief account will suffice for the applications we shall make of this lemma. We refer the reader to the literature for details on other forms of Zorn's lemma, and their equivalence to the well-ordering axiom or the axiom of choice.

5. DIVISIBLE GROUPS

In an abelian group any element may be multiplied by an integer. But what about dividing by an integer? The answer is that the result may not exist, and if it exists, it may not be unique. So we shall not attempt to attach a meaning to the symbol $\frac{1}{n}x$, but nevertheless we shall say that x is divisible by n if there exists y with $ny = x$.

Examples. (*a*) The element 0 is divisible by any integer.

(*b*) If x has order m, then it is divisible by any integer prime to m.

(*c*) In the additive group of rational numbers, every element is divisible by every integer.

In this section we are going to study groups which share this last property with the additive group of rational numbers.

Definition. A group G is *divisible* if for every x in G and every integer n there exists an element y in G with $ny = x$.

Alternatively, G is divisible if $G = nG$ for every integer n.

We note that a cyclic group is not divisible. Nor for that matter is a direct sum of cyclic groups. Indeed, it is clear that *a direct sum of groups is divisible if and only if every summand is divisible*. Another easily verified fact is that *a homomorphic image of a divisible group is divisible*. So we note that the group of rationals mod one is divisible, since it is a homomorphic image of the additive group of rationals.

The group $Z(p^\infty)$ is also divisible. This is not apparent from the definition (§ 2) of $Z(p^\infty)$ as P/Z, since P (the group of rationals with denominator a power of p) is not divisible. If we admit, as was claimed at the end of § 3, that $Z(p^\infty)$ is a direct summand of the rationals mod one, then the divisibility of $Z(p^\infty)$ is assured. But let us give a direct argument. Since $Z(p^\infty)$ is a primary group, all of its elements are divisible by any integer prime to p. On the other hand, it is clear that every element of $Z(p^\infty)$ can be divided by arbitrary powers of p. On putting these two statements together, we establish that $Z(p^\infty)$ is divisible.

The theory of divisible groups is based on the theorem below. It is to be understood that by a "divisible subgroup" we mean a subgroup which as a group on its own merits is divisible. In other words, for H to be a divisible subgroup of G, it has to be the case that for every $h \in H$, and every integer n, there exists an element h_1 *again in* H, satisfying $nh_1 = h$.

Theorem 2. A divisible subgroup of an abelian group is a direct summand.

Proof. Let H be a divisible subgroup of G. Our task is to find a subgroup K with $H \cap K = 0$, $H + K = G$. Offhand, it probably seems difficult to imagine how to go about finding such a subgroup. It is rather remarkable that a crude use of Zorn's lemma accomplishes the objective.

We consider the set $\mathfrak{S}$ of all subgroups L which satisfy $H \cap L = 0$. (There is at least one, namely, 0.) We would like to get one as large as possible. So we set out to get a maximal element in $\mathfrak{S}$. We partially order $\mathfrak{S}$ by set-theoretic inclusion. To use Zorn's lemma, we have to verify that every chain in $\mathfrak{S}$ has a least upper bound. Suppose $\{L_i\}$ is a chain in $\mathfrak{S}$. To get the desired least upper bound, we simply take the set-theoretic union of the L_i's, say M. Three things need to be verified:

(*a*) M is a subgroup. We take x and y in M and have to show that $x - y$ is in M. Now x and y got into M only because x was, say, in L_i, y in L_j. Moreover, L_i and L_j are comparable, say $L_i \subseteq L_j$. Then *both* x and y are in L_j, and so is $x - y$. Hence $x - y$ is in M.

(*b*) $H \cap M = 0$. This follows from the fact that every element of M is in one of the L_i's, and each $H \cap L_i = 0$.

(*c*) M is the least upper bound of $\{L_i\}$. This is clear.

We remark that the arguments above are of a routine nature. Indeed the whole would usually be condensed to: "By Zorn's lemma pick a subgroup maximal with respect to disjointness from H." In the future we shall give such a condensed version. But the reader should observe that there is one vital point which must be checked before Zorn's lemma is applicable—that the property of disjointness from H is preserved under taking of least upper bounds of chains.

At any rate, we now have a maximal subgroup K in $\mathfrak{S}$, and we set out to prove $H + K = G$. We suppose the contrary. Then there exists an element x which is not in $H + K$. A fortiori, x is not in K. We now form K', the subgroup generated by K and x. K' is larger than K, and, in fact, K' consists of all elements $k + nx$ where k is in K and n is an integer. By the maximality of K we know that $H \cap K' \neq 0$. Hence there exists a nonzero element h in $H \cap K'$:

$$h = k + nx. \tag{3}$$

From equation (3) we see that nx is in $H + K$. It is interesting to observe that we have not yet used the divisibility of H. In other words, we have proved that if we take any subgroup H and a maximal subgroup

K disjoint from it, then $H + K$ is at any rate large enough so that $G/(H + K)$ is a torsion group.

Now to complete the proof. We may suppose that n is the smallest positive integer such that $nx \in H + K$ (of course $n > 1$). Let p be a prime dividing n, and write $y = (n/p)x$. Thus y is not in $H + K$, but $py = nx = h - k$. By the divisibility of H we may write $h = ph_1$, with $h_1 \in H$. Let $z = y - h_1$. Then z is not in $H + K$, but $pz = -k$ is in K. We now repeat the argument above; when we adjoin z to K we must get a subgroup not disjoint from H. Hence we have

$$h_2 = k_2 + mz \tag{4}$$

where $h_2 \in H$, $h_2 \neq 0$, $k_2 \in K$, and m is an integer. It is impossible that m be a multiple of p, for then the right side of equation (4) is in K, while the left side is a nonzero element of H. Hence m is prime to p, and we may find integers a, b such that $am + bp = 1$. We have $z = amz + bpz \in H + K$, a contradiction. This completes the proof of Theorem 2.

It is worth while to take another glance at the mechanism of the preceding proof. A transfinite induction contains two steps: a passage to the limit, and an argument for pushing one stage further. The first step was accomplished above by the initial application of Zorn's lemma. The second was concealed in an indirect proof, but it is perfectly possible to rewrite this as a direct proof. It is rather typical of the use of Zorn's lemma that it culminates in an indirect argument.

We proceed to a useful theorem applying to an arbitrary group G. In G consider the totality of divisible subgroups and form their union M (this is one of the rare occasions when we could correctly construe M to be the set-theoretic rather than the group-theoretic sum, for there is actually a largest subgroup among them). Now M consists of finite sums $x_1 + \cdots + x_k$ where each x_i lies in some divisible subgroup. Since each x_i is divisible by n (for arbitrary n), so is the sum. Thus M is itself a divisible group. We have proved the first statement of the following theorem:

Theorem 3. Any abelian group G *has a unique largest divisible subgroup* M, *and* $G = M \oplus N$ *where* N *has no divisible subgroups.*

To prove the last statement of Theorem 3 we quote Theorem 2 and deduce that M is a direct summand of G. The other summand N can have no divisible subgroups, for these would be divisible subgroups of G.

The subgroup M is uniquely determined, for it is intrinsically characterized as the maximal divisible subgroup. Suitable examples show that N is not necessarily unique. Of course, N is unique *up to isomorphism*, for it is isomorphic to G/M.

Theorem 3 suggests the following definition:

Definition. An abelian group is *reduced* if it has no (nonzero) divisible subgroups.

To classify all abelian groups it suffices by Theorem 3 to do the divisible and reduced cases. We proceed at once to a complete determination of divisible groups. After this has been done, it will usually be possible to restrict our attention to reduced groups.

Theorem 4. A divisible abelian group is a direct sum of groups each isomorphic to the additive group of rational numbers or to $Z(p^\infty)$ (*for various primes* p).

Proof. Let G be the group, T its torsion subgroup. It is easy to verify that T is again divisible. By Theorem 2, $G = T \oplus F$ where F is isomorphic to G/T, and so is divisible and torsion-free. We now study T and F separately.

The discussion of F is easy to carry out directly, but it will clarify the situation to relate it to standard vector space theory. Let x be any element in F, and n a nonzero integer. Then, since F is divisible and torsion-free, there is *exactly one* element y in F with $ny = x$. Thus we can attach a unique meaning to $\left(\frac{1}{n}\right)x$, and then to rx, where r is a rational number. Now it is routine to check the requisite postulates, and we conclude that F is a vector space over the field of rational numbers. Any vector space (whether finite or infinite-dimensional) has a basis. Translated into group-theoretic terms, this says that F is a direct sum of groups each isomorphic to the additive group of rational numbers.

We turn our attention now to the divisible torsion group T. By Theorem 1, T is a direct sum of primary groups, each of which will again be divisible. So we may as well assume that T itself is a primary group (say for the prime p), and we have to prove that T is a direct sum of groups isomorphic to $Z(p^\infty)$.

A little care must be exercised in applying Zorn's lemma for this purpose. We consider subgroups of T isomorphic to $Z(p^\infty)$. Of course it is not clear at the moment that any such subgroups exist, but this is something to worry about later. Since the objective is to express T as a direct sum of such subgroups, it is appropriate to consider independent sets of these subgroups. So we decide to form $\mathfrak{P}$, the set of all independent sets of subgroups isomorphic to $Z(p^\infty)$. It should be borne in mind that each element of $\mathfrak{P}$ is an independent set of subgroups, that is to say, a set of sets; so $\mathfrak{P}$ is a set of sets of sets! We introduce in $\mathfrak{P}$ the natural ordering given by set-theoretic inclusion. The proof that every chain in $\mathfrak{P}$ has a least upper bound offers no difficulty. Thus we may apply Zorn's lemma to arrive at a maximal independent set of subgroups isomorphic to $Z(p^\infty)$, say $\{S_i\}$. Write $S = \Sigma S_i$. The proof will be finished if we show $S = T$. In any event, S is a divisible group (being a direct sum of divisible groups), and so $T = S \oplus R$ by Theorem 2. Now we come to the crux of the proof. If $R \neq 0$, we shall show that

R contains a subgroup isomorphic to $Z(p^\infty)$; by adjoining this subgroup to $\{S_i\}$ we get a contradiction, for the enlarged set of subgroups is still independent.

We select in R an element x_1 of order p. Using the divisibility of R, we find in succession elements $x_2, x_3, \cdots$ with $px_2 = x_1$, $px_3 = x_2, \cdots$, and in general $px_{i+1} = x_i$. Now map x_1 into $1/p$, x_2 into $1/p^2, \cdots, x_i$ into $1/p^i, \cdots$. This gives rise to an isomorphism between the subgroup generated by the x's and the group $Z(p^\infty)$, and completes the proof of Theorem 4.

We conclude this section with two remarks:

(*a*) The particular way in which Zorn's lemma was applied deserves comment. It would perhaps have seemed more natural to consider all subgroups which are direct sums of groups isomorphic to $Z(p^\infty)$, and proceed to use Zorn's lemma to get a maximal one. But there is a catch: Why should the union of an ascending chain of these subgroups be expressible as a direct sum of $Z(p^\infty)$'s? It is true that each subgroup in the chain will have such an expression, but these expressions will presumably be unrelated, and it is impossible to combine them. It was to obviate this difficulty that we chose the fussier formulation above.

(*b*) The uniqueness question that arises in connection with Theorem 4 also warrants attention. We are presented with a set of cardinal numbers: one for the number of rational summands, and then one for every p giving the number of $Z(p^\infty)$ summands. It is a fact that these cardinal numbers are invariants (and consequently, of course, a complete set of invariants). For the rational summands this is simply a restatement of the invariance of the number of elements in a basis of a vector space, a fact which we shall take for granted. For the $Z(p^\infty)$ summands the question of uniqueness of the cardinal number can be rapidly reduced to the case of a vector space: all we have to do is drop down to the subgroup of all elements x satisfying $px = 0$ (this being a vector space over the field of integers mod p).

For an alternative approach to Theorems 2 and 4 the reader is referred to exercises 1-4.

Exercises 1-8

1. Let G be a group, H a subgroup, D a divisible group. Let f be a homomorphism of H into D. Show that f can be extended to a homomorphism of G into D. (First study the task of extending f to the subgroup of G generated by H and one more element; it turns out that the divisibility of D always makes this possible. It remains to prepare the way for application of Zorn's lemma. This may be done as follows: Consider pairs (S_i, f_i), where S_i is a subgroup of G containing H, and f_i is an extension of f. Partially order these pairs by decreeing that $(S_i, f_i) \geq (S_j, f_j)$ means that $S_i \supseteq S_j$ and that f_i is an extension of f_j. Apply Zorn's lemma.)

2. Let G be a group, H a subgroup, and suppose that the identity homomorphism of H onto itself can be extended to a homomorphism of G onto H. Prove that H is a direct summand of G. (If K is the kernel of the extension, then K is the desired complementary summand to H.)

3. Combine exercises 1 and 2 to get another proof of Theorem 2. (This proof is in a way more conceptual and therefore more satisfying than the one in the text. But the latter was peculiarly suitable for the first illustration of Zorn's lemma.)

4. Prove Theorem 4 as follows: Let T be a divisible primary group. Let P be the subgroup of elements y satisfying $py = 0$; choose a basis of P, thinking of P as a vector space over the integers mod p. If $x = x_1$ is a basis element, write $x_1 = px_2 = p^2x_3$, etc. Thus we get, for each basis element, a subgroup isomorphic to $Z(p^\infty)$, and T is their direct sum.

5. Prove that any group can be embedded in a divisible group. (Write G as F/H, where F is free. Embed F in a direct sum of copies of the rational numbers, say E. Then $G \subset E/H$ is the desired embedding.)

6. Prove that if a group is a direct summand of every containing group it is divisible.

7. Prove that the following three properties of a group are equivalent: (*a*) G is divisible; (*b*) G is a direct summand of every containing group; (*c*) if K is a group and f a homomorphism of a subgroup of K into G, then f can be extended to a homomorphism of K into G.

8. Let G be a group, H a subgroup. Assume that G/H is divisible and torsion-free, and that H is of bounded order ($nH = 0$ for some n). Prove that H is a direct summand of G. (The desired complement is nG; in fact it is the only complement.)

6. TWO TEST PROBLEMS

We interrupt the exposition at this point to raise a (largely rhetorical) question: What is our goal in studying abelian groups? There is an obvious, optimistic answer which can be phrased in various ways: We seek to classify all abelian groups, or to give a complete set of invariants for abelian groups, or to give necessary and sufficient conditions for two abelian groups to be isomorphic.

At present there is no visible prospect of doing anything like this for general abelian groups. But we do expect to achieve structure theorems for certain restricted classes of groups—one of them (Theorem 4) is already in our possession.

However, in connection with these theorems it is wise to sound a note of caution. How do we know when we have a satisfactory theorem? After all, a "complete set of invariants" might turn out to be so complicated as to be virtually useless. We suggest that a tangible criterion be employed: the success of the alleged structure theorem in solving an explicit problem. The following problems are proposed for the purpose.

Problem I. If G is isomorphic to a direct summand of H, and H is isomorphic to a direct summand of G, are G and H necessarily isomorphic?

Problem II. If $G \oplus G$ and $H \oplus H$ are isomorphic, are G and H isomorphic?

A few remarks on these problems are in order:

(*a*) Both problems can be formulated for very general mathematical systems, and only rarely is the answer known. Note in particular that Problem I has an affirmative answer in set theory, namely the Cantor-Schröder-Bernstein theorem.

(*b*) The reader may wonder why in Problem I we did not merely assume that each group is a subgroup of the other (instead of a direct summand). The reason is that in this stronger version the problem has an easy negative answer. For example, take

$$G = Z_2 \oplus Z_4 \oplus Z_4 \oplus Z_4 \oplus \cdots$$

$$H = Z_4 \oplus Z_4 \oplus Z_4 \oplus \cdots$$

That G and H are not isomorphic follows from the uniqueness of the representation of a primary group as a direct sum of cyclic groups (see § 11). But each is isomorphic to a subgroup of the other.

(*c*) Problem II can be tested on finite groups. We see that the mere knowledge that a finite group is a direct sum of cyclic groups is not enough; we have to have a uniqueness statement to go with it.

(*d*) Both problems have affirmative answers for divisible groups (exercise 9).

(*e*) By a sophisticated structure theory, we shall find that both problems have affirmative answers for countable torsion groups (exercise 32).

Exercises 9-10

9. Show that the answer to both problems is affirmative for divisible groups. (Note that in solving Problem I the ordinary Cantor-Schröder-Bernstein theorem for cardinal numbers is needed.)

10. Show that both problems would be defeated by the existence of groups G, H such that G is isomorphic to $G \oplus H \oplus H$ but not to $G \oplus H$.

7. PURE SUBGROUPS

The concept we are about to introduce is attributable to Prüfer [102]; he used for it the term "Servanzuntergruppe," which has been translated into English in various ways. We shall, however, follow Braconnier [15] in using the designation "pure."

Definition. A subgroup H of an abelian group G is *pure* if $h \in H$, $h = ny$ (n an integer, $y \in G$) imply $h = nh_1$ with h_1 in H. In other words, if an element of H is divisible by n in G, it is already divisible by n in H.

The following remarks may help the reader to digest this new idea (the proofs are easy, and we omit most of them):

(*a*) A pure subgroup of a pure subgroup is pure.

(*b*) A divisible subgroup is pure. (One might think of purity as a kind of relative divisibility. For H to be divisible, its elements have to be divisible by arbitrary integers. For H to be pure, its elements have to admit division by integers only when such division is possible in G.)

(*c*) A subgroup of a divisible group is pure if and only if it is divisible.

(*d*) The torsion subgroup of a group is pure. (It even satisfies the stronger condition of being *closed* under division by integers, whenever such division is possible.)

(*e*) If S is a subgroup of G such that G/S is torsion-free, then S is pure. (In fact, this condition on S is equivalent to the stronger version of purity just referred to.)

(*f*) If G itself is torsion-free, purity of S simply means that S is closed (within G) under division by integers, the latter being now unique. From this one draws two corollaries, valid only in this torsion-free case: any intersection of pure subgroups is pure; any subgroup is contained in a unique smallest pure subgroup.

(*g*) Here is the simplest example of a nonpure subgroup: Let G be the additive group of integers mod 4, $G = \{0, 1, 2, 3\}$, and $H = \{0, 2\}$. Then 2 is a multiple of 2 in G but not in H, so H is not pure.

(*h*) Any direct summand is pure. Suppose $G = S \oplus T$, $x \in S$, and $x = ny$ with $y \in G$. Then $x = ny_1$, where y_1 is the component of y in S. Hence S is pure.

(*i*) A pure subgroup need not be a direct summand. In fact (exercise 33) there exists a group for which even the torsion subgroup is not a direct summand. On the other hand, Theorems 5 and 7 below give two special cases where it is true that pure subgroups are direct summands.

(*j*) The union of an ascending chain of pure subgroups is pure. It is this property that accounts for the importance of the concept of purity. The union of an ascending chain of direct summands is not necessarily a direct summand. For this reason, although it is direct summands that we are mostly interested in, where an application of Zorn's lemma is impending pure subgroups will be decidedly more convenient to use.

The following lemma on pure subgroups is basic:

Lemma 1. Let G *be an abelian group,* H *a pure subgroup, and* y *an element of* G/H. *Then there exists an element in* G, *mapping on* y *mod* H, *and having the same order as* y.

Proof. If y has infinite order, then any choice of an element mapping on y will do. So suppose y has finite order n. First choose any z in G mapping on y. Then nz is in H. By the purity of H, there exists an element $h \in H$ with $nz = nh$. Set $x = z - h$. Then x has the desired properties: it maps on y mod H, and has order n.

From Lemma 1 we readily deduce our first theorem asserting that a pure subgroup is sometimes a direct summand:

Theorem 5. Let G *be an abelian group and* H *a pure subgroup such that* G/H *is a direct sum of cyclic groups. Then* H *is a direct summand of* G.

Proof. For each cyclic summand of G/H pick a generator y_i. By Lemma 1, we select elements x_i in G, mapping on y_i mod H, and having the same order. (The careful reader will observe that an application of the axiom of choice in its original Zermelo version is hidden in this innocent sentence.) Let K be the subgroup of G generated by the elements x_i. We claim that $G = H \oplus K$.

(*a*) $H + K = G$. Let t be any element in G, mapping let us say on t^* in G/H. We may write t^* as a finite sum $\Sigma a_i y_i$, with integral coefficients. Then $t - \Sigma a_i x_i$ maps on 0 in G/H, and so lies in H. Since $\Sigma a_i x_i \in K$, we have $t \in H + K$.

(*b*) $H \cap K = 0$. Let $w \in H \cap K$, say $w = \Sigma a_i x_i$. Since w is in H, we have $\Sigma a_i y_i = 0$. If y_i has infinite order, this means $a_i = 0$; if y_i has finite order n_i, then a_i must be a multiple of n_i. In either case, $a_i x_i = 0$, $w = 0$.

We shall conclude this section with two lemmas (to be used later) concerning the behavior of purity with respect to homomorphism. (See exercise 13 for more on this.)

Lemma 2. Let G *be a group,* S *a pure subgroup of* G, *and* T *a subgroup containing* S *such that* T/S *is pure in* G/S. *Then* T *is pure in* G.

Proof. Suppose $t \in T$ and $t = nx$ with $x \in G$. We have to prove that t is a multiple of n in T. Let t^* and x^* be the homomorphic images of t and x in G/S. Then $t^* = nx^*$, and by the purity of T/S, $t^* = ny$ with $y \in T/S$. Let t_1 be an element in T mapping on y mod S. Then $t = nt_1 + s$ for a suitable element $s \in S$. Since $s = t - nt_1 = nx - nt_1$, and S is pure, we conclude that $s = ns_1$ with s_1 in S. This gives us $t = n(t_1 + s_1)$, as desired.

Lemma 3. Let S *be a pure subgroup of* G *with* $nS = 0$. *Then* $(S + nG)/nG$ *is pure in* G/nG.

Proof. Suppose $x = my$ where $x \in (S + nG)/nG$, $y \in G/nG$, and m is an integer. We have to prove that x is a multiple of m within $(S + nG)/nG$. Let us take representatives s, t in G of x and y; the representative s of x may evidently be chosen in S. Then s and mt differ by an element of nG:

$$s = mt + nz \qquad (z \in G). \tag{5}$$

Write r for the greatest common divisor of m and n. Then divide m and n by r, to get $m = rm_1$, $n = rn_1$. Since m_1 and n_1 are relatively prime, we can find integers a and b such that $am_1 + bn_1 = 1$. Equation (5) shows that s is a multiple of r in G. Since S is pure, $s = rs_1$ with $s_1 \in S$. Hence

$$s = rs_1 = r(am_1 + bn_1)s_1 = mas_1 + nbs_1 = mas_1, \tag{6}$$

the last step following from $ns_1 \in nS = 0$. After passage modulo nG, equation (6) shows that x is a multiple of m within $(S + nG)/nG$, as desired.

Exercises 11-13

11. Give examples to show that neither the union nor the intersection of pure subgroups need be pure.

12. For what groups is it true that every subgroup is pure? (Answer: Torsion groups in which every element has square-free order. In other words, a direct sum of vector spaces over the integers mod p.)

13. (*a*) Let S and T be pure subgroups of G, with $S \subset T$. Prove that T/S is pure in G/S. (This is the converse of Lemma 2.)

(*b*) Give an example to show that we cannot dispense with the hypothesis $S \subset T$. In other words, purity of S and T need not imply purity of $(S + T)/S$ in G/S. (But observe that Lemma 3 is a special case where this does hold.)

8. GROUPS OF BOUNDED ORDER

We say that a group is of ***bounded order*** if it is, in the first place, a torsion group, so that all elements have finite order, and if, further, there is a fixed upper bound to the orders of the elements. In other words, there is to exist a (positive) integer n such that $nx = 0$ for all x, or, more briefly still, such that $nG = 0$. Of course any finite group is of bounded order. But an infinite group can also be of bounded order: take the direct sum of an infinite number of finite cyclic groups, having an upper bound on the orders of the summands.

We shall shortly see that any group of bounded order is a direct sum of cyclic groups. In a way, this is the most satisfactory generalization of the theorem that a finite group is a direct sum of cyclic groups.

In motivating the next lemma we remark that, a priori, there seems to be no visible means of constructing even a single cyclic direct summand in a given group of bounded order. But Lemma 4 shows how easy it is to obtain a pure cyclic subgroup. This clearly illustrates the advantage of a pure subgroup as a temporary substitute for a direct summand.

Lemma 4. Let G *be a primary group satisfying* $p^r G = 0$. *Let* x *be an element of order* p^r *in* G. *Then the cyclic subgroup* K *generated by* x *is pure.*

Proof. Let us begin by recalling that in a primary group every element is automatically divisible by any integer prime to p. From this we see that in checking purity we need only concern ourselves with powers of p; also, in looking at multiples of x we can confine ourselves to the elements $p^i x$. In the future we shall often make tacit use of these simplifications.

Suppose then that $p^i x = p^j y$ for $i < r$ and some y in G. We have to prove that $p^i x$ is a multiple of p^j within K. If $j \leqq i$, this is of course clear. If $j > i$, we have $0 = p^r y = p^{r-j}(p^i x)$, contradicting the hypothesis that x has order p^r.

We pause at this point to observe that we have accumulated enough information to settle the case of a finite group G. For Lemma 4 gives us a pure cyclic subgroup K; by induction G/K is a direct sum of cyclic groups; then, by Theorem 5, K is a direct summand of G.

If we try to imitate this procedure in the infinite case, we are impeded by the fact that no inductive assumption can tell us that G/K is a direct sum of cyclic groups. Instead we shall build up pure subgroups which are direct sums of cyclic groups. In order to insure the necessary independence of subgroups, we insert at this point an elementary lemma.

Lemma 5. Let G *be a group,* S *a subgroup,* x *an element of* G *mapping on* y *mod* S. *Suppose that* x *and* y *have the same order. Let* K *be the cyclic subgroup generated by* x. *Then the union* S + K *is a direct sum.*

Proof. We have merely to show that $S \cap K = 0$. Suppose rx is in $S \cap K$. Then ry is necessarily 0, and r is a multiple of the order of y, whence rx = 0 by hypothesis.

Theorem 6. A group of bounded order is a direct sum of cyclic groups.

Proof. We can assume (Theorem 1) that G is primary. The method of proof is to build up a direct sum of cyclic groups, preserving the purity of the sum as we go along. In the interest of brevity, it will pay us to define a subset of G to be *pure* if it generates a pure subgroup. We focus attention on *pure independent* subsets of G; and when we do so the operation becomes perfectly analogous to finding a basis of a vector space. In fact, we select (Zorn's lemma) a maximal pure independent set $\{x_i\}$. If the subgroup S generated by the elements x_i is all of G, we are finished. Otherwise, we observe that G/S is again primary of bounded order and we select an element y in G/S of maximal order. By Lemma 4 the cyclic subgroup generated by y is pure in G/S. By Lemma 1 we may find x in G, mapping on y mod S, and having the same order as y. Lemmas 2 and 5 now apply precisely to show that the enlarged set $\{x, x_i\}$ is still pure and independent. This contradiction shows that S is indeed all of G, and concludes the proof of Theorem 6.

The question of the uniqueness of the decomposition is of almost equal interest; but it is convenient to postpone the discussion of uniqueness to § 11.

With the aid of Theorem 6 we shall next prove a second theorem concerning the connection between pure subgroups and direct summands. We need a preliminary lemma which is really lattice-theoretic in nature.

Lemma 6. Let S *and* T *be subgroups of* G *with* $S \cap T = 0$, *and suppose that* $(S + T)/T$ *is a direct summand of* G/T. *Then* S *is a direct summand of* G.

Proof. Let R/T be a complementary summand to (S + T)/T in G/T. We have $R + (S + T) = G$, $R \cap (S + T) = T$. We assert that $G = S \oplus R$. For since $R \supset T$, we have $S + R = S + T + R = G$. Again $(R \cap S) \subset R \cap (S + T) = T$, and hence $R \cap S \subset T \cap S = 0$.

Theorem 7. Let G *be a group and* S *a pure subgroup of bounded order. Then* S *is a direct summand of* G.

Proof. Suppose $nS = 0$. Then, by Lemma 3, (S + nG)/nG is pure in G/nG. Also, G/nG and all its homomorphic images are groups of bounded order; hence by Theorem 6 they are direct sums of cyclic groups. By Theorem 5, (S + nG)/nG is a direct summand of G/nG. We next note that $S \cap nG = 0$; this is an immediate consequence of the purity of S and the fact that $nS = 0$. We are ready to apply Lemma 6 (with nG playing the role of T), and we deduce that S is a direct summand of G.

Since the torsion subgroup is always pure we see, as a special case of Theorem 7, that the torsion subgroup is a direct summand if it is of bounded order. With the aid of Theorem 2 we can go a little further:

Theorem 8. Let G *be a group,* T *its torsion subgroup. Suppose that* T *is the direct sum of a divisible group and a group of bounded order. Then* T *is a direct summand of* G.

Exercises 14-19

14. Prove Theorem 6 as follows: Let G be primary, let $p^rG = 0$, and let P be the subgroup of elements with $px = 0$. Choose a basis for $P \cap p^iG$ modulo $P \cap p^{i+1}G$; express every such basis element as a multiple of p^i. Gather up the resulting elements $(i = 0, \cdots, r - 1)$, and you have a basis of G. (This is probably the most direct road to Theorem 6. The method in the text has, however, expository merits in that the side results are instructive.)

15. Let G be a group of bounded order, S a pure subgroup. Prove that S is a direct summand by taking a maximal pure independent set in S and expanding it to one in G. (In this way one can prove Theorem 7 without using Theorem 5.)

16. Let G be the direct sum of cyclic groups of order p^i generated by x_i $(i = 1, 2, \cdots)$. Let S be the subgroup generated by all $x_i - px_{i+1}$. Show that: (*a*) S is pure, (*b*) G/S is isomorphic to $Z(p^\infty)$, (*c*) S is not a direct summand of G.

17. Let G be a primary group such that every pure subgroup is a direct summand. Prove that G is the direct sum of a divisible group and a group of bounded order. (After the maximal divisible subgroup has been disposed of, it can be assumed that G is reduced. Then show that G is a direct sum of cyclic groups. If G is not of bounded order, generalize exercise 16.)

18. (*a*) Let G be primary of bounded order, and P the subgroup of elements x with $px = 0$. Let Q be a subgroup of P. Show that there exists a direct summand H of G such that $H \cap P = Q$.

(*b*) Is this true if G is not of bounded order?

19. Let G be a primary group. Show that G admits a homomorphism onto $Z(p^\infty)$ if and only if it is not of bounded order.

9. HEIGHT

On several previous occasions we have discussed the extent to which a given element in a group can be divided by integers. In this section a numerical measure of this divisibility will be introduced. We are concerned only with the case of a primary group, but the concept is definable in any group (one would then speak of the height at p for each prime p).

Definition. Let G be a primary group, for the prime p. An element x in G has ***height*** n if x is divisible by p^n but not by p^{n+1}; it has ***infinite height*** if it is divisible by p^n for every n.

We write h(x) for the height of x; thus h(x) is a (nonnegative) integer or the symbol ∞.

If x lies in a subgroup S of G, we may define two heights for x. When it is necessary to make a distinction, we shall write $h_S(x)$ and $h_G(x)$ for the height of x in S and G, respectively. Note that we always have $h_S(x) \leqq h_G(x)$.

Remarks. (*a*) If h(x) and h(y) are unequal, then h(x + y) is precisely the smaller of the two. If h(x) = h(y), then $h(x + y) \geqq h(x)$ and may even be larger. We shall make constant use of this observation in the future.

(*b*) $h(0) = \infty$. As a rule we take this trivial fact for granted and disregard it; thus when we say that G has no elements of infinite height, we mean no elements other than 0.

(*c*) The element x has height n if and only if x is in p^nG but not in $p^{n+1}G$. Thus we have a descending chain of subgroups

$$G \supset pG \supset p^2G \supset \cdots \supset p^nG \supset \cdots,$$

and the elements of height n are those which drop out in the passage from p^nG to $p^{n+1}G$. The elements of infinite height comprise the intersection of all p^nG.

(*d*) G is divisible if and only if all its elements have infinite height.

(*e*) The elements of infinite height form a subgroup H. It is fallacious to argue as follows: By remark (*d*), H is divisible; then by Theorem 2, H is a direct summand of G. The mistake is that the elements of H, while they have infinite height in G, need not have infinite height in H. It is a fact that H need not be a direct summand (see in this connection exercise 30).

(*f*) In the group $Z(p^\infty)$ the element 1/p sits on a descending chain: $1/p, 1/p^2, 1/p^3, \cdots$. Such an infinite descending chain of course shows that 1/p has infinite height. But it should not be supposed that an element of infinite height necessarily has such a chain under it; it may instead have various finite chains of ever greater length.

(*g*) It is clear that a *necessary* condition for G's being a direct sum of cyclic groups is that it have no elements of infinite height. It is a remarkable fact that the converse holds if G is countable, but fails otherwise (Theorem 11 and exercise 33).

(*h*) Since in a primary group only divisibility by powers of p matters, we may recast the definition of purity as follows: A subgroup S of G is pure if and only if its elements have the same height in S as in G. In the next lemma we refine this remark by showing that we can even confine ourselves to the elements of order p. (Note that Lemma 4, above, is in essence a special case of Lemma 7.)

Lemma 7. Let G *be a primary group and* S *a subgroup with no elements of infinite height. Suppose that the elements of order* p *in* S *have the same height in* S *as in* G. *Then* S *is pure.*

Proof. Given x in S, we have to prove $h_S(x) = h_G(x)$. We may assume by induction that this is known for elements of order $\leqq p^n$, and we suppose that x has order p^{n+1}. Suppose $h(px) = r$; we observe that this notation is unambiguous, since the height of px is the same in S as in G. Write $px = p^r y$ with y in S. Then the height of $p^{r-1}y$ must be precisely $r - 1$ in both S and G; for if it were larger (even in G), then $h(p^r y)$ would exceed r. Considering the height of x, in S or G, we can at any rate observe that it does not exceed $r - 1$. Now we write

$$x = (x - p^{r-1}y) + p^{r-1}y.$$

Since $p(x - p^{r-1}y) = 0$, the element $x - p^{r-1}y$ has an unambiguous height, say k. If $k \neq r - 1$, then $h_S(x) = h_G(x)$, since both must be precisely the minimum of k and $r - 1$. If $k = r - 1$, then both heights are at least $r - 1$. But, on the other hand, they do not exceed $r - 1$, as we noted above. Hence they are both equal to $r - 1$.

The next lemma is another instance of the phenomenon that the elements of order p somewhat determine the behavior of all elements.

Lemma 8. Let G *be a primary group and suppose that all elements of order* p *in* G *have infinite height. Then* G *is divisible.*

Proof. We have to prove that every x in G has infinite height. Suppose, on the contrary, that $h(x) = m < \infty$. By induction on the order we have $h(px) = \infty$, and so we may write $px = py$ with $h(y) > m$. Then $h(x - y)$ is, on the one hand, m (the smaller of the two heights), and, on the other hand, ∞, since $p(x - y) = 0$.

We are now able to prove a fairly far-reaching theorem on the existence of cyclic direct summands:

Theorem 9. Let G *be a reduced group which is not torsion-free. Then* G *has a finite cyclic direct summand.*

Proof. First we take up the case where G is primary. By hypothesis, G is not divisible. Hence by Lemma 8 there exists in G an element x of order p and finite height. Say $h(x) = m$, write $x = p^m y$, and let H be the cyclic subgroup generated by y. It follows from Lemma 7 that H is pure in G (note that the only elements of order p in H are the multiples of x by integers prime to p). By Theorem 7, H is a direct summand of G.

Now let G be arbitrary. By hypothesis, its torsion subgroup T is nonzero. Let S be a nonzero primary summand of T. Then S cannot be divisible, and we have just seen that S has a finite cyclic direct summand H. Now H is pure in S, which is pure in T, which is pure in G. By these three successive steps we see that H is pure in G. By Theorem 7, H is a direct summand of G. This completes the proof of Theorem 9.

We define a group G to be *indecomposable* if it cannot be written as a direct sum, except in the trivial way $G \oplus 0$.

Theorem 10. An indecomposable group cannot be mixed; that is, it is either a torsion group or torsion-free. If it is a torsion group, it is isomorphic to $Z(p^n)$ *or* $Z(p^\infty)$ *for some prime* p.

Proof. If the given group G is torsion-free, there is nothing to prove. So we suppose the contrary. Then if G is divisible, it is isomorphic to $Z(p^\infty)$ by Theorem 4. So we may assume that G is reduced. Theorem 9 then completes the proof.

We thus have determined all indecomposable torsion groups, and found there are none other than the familiar ones. But it is quite another matter, and a decidedly unsolved problem, to determine the indecomposable torsion-free groups. We shall discuss this subject in § 15.

Exercises 20-27

20. Let G be a primary group with no elements of infinite height. Show that G can be embedded in a complete direct sum of cyclic groups. (Embed G in the complete direct sum of the groups G/p^nG.)

21. Prove that a group has the descending-chain condition on subgroups if and only if it is the direct sum of a finite group and a finite number of copies of groups isomorphic to $Z(p^\infty)$, for various primes p. (Note that for torsion groups the ascending-chain condition implies finiteness, and so is stronger than the descending-chain condition.)

22. Let G be a primary group and S a subgroup generated by a pure independent set $\{x_i\}$. Prove that $\{x_i\}$ is maximal if and only if G/S is divisible.

23. Let G be an infinite group such that every proper subgroup is finite. Prove that G is isomorphic to $Z(p^\infty)$.

24. Let G be an infinite group such that every proper homomorphic image is finite. Prove that G is cyclic.

25. Let G be an infinite group which is isomorphic to every proper subgroup. Prove that G is cyclic.

26. Let G be an infinite group which is isomorphic to every proper homomorphic image. Prove that G is isomorphic to $Z(p^\infty)$.

27. Let G be a reduced primary group which is not of bounded order. Prove that G has cyclic direct summands of arbitrarily high order.

10. DIRECT SUMS OF CYCLIC GROUPS

Let us consider how one would go about using Theorem 9 to get a structure theorem. For instance, consider a reduced primary group G. By Theorem 9, G has a cyclic direct summand, say H_1. By a second application of Theorem 9, the other summand likewise has a cyclic direct summand, H_2. We may continue in this way to detach cyclic summands. Does this prove that G is a direct sum of cyclic groups? It cannot, since there exist plenty of reduced primary groups which are not direct sums of cyclic groups. The trouble is that we have no assurance that the sequence of the H's will use up all of G; and there is no prospect of continuing by transfinite induction, since the union of the H's is presumably not a direct summand.

If we assume that G is countable, then there will be at least a chance for the sequence of H's to exhaust G, *if it is skillfully selected.* Theorem 11 shows that a suitable sequence can indeed be selected, if we make the hypothesis (in any case necessary) that G has no elements of infinite height. Theorem 11 is a special case of Theorems 12 and 14 below, but it is of some interest to give an independent proof of this more elementary theorem.

We need a preliminary lemma for which countability is irrelevant:

Lemma 9. Let G *be a primary group with no elements of infinite height,* H *a finite pure subgroup,* x *any element of* G. *Then there exists a finite pure subgroup of* G *containing* H *and* x.

Proof. By Theorem 7, $G = H \oplus K$. Suppose $x = y + z$ in this decomposition and $p^r z = 0$. We make an induction on r, and we may thus assume that H and pz can be embedded in a finite pure subgroup H'. In the decomposition $G = H' \oplus K'$, let $z = v + w$. Since $pz \in H'$, we have $pw = 0$. By hypothesis $h(w) = m$ is finite. Write $w = p^m u$. By Lemma 7 the cyclic subgroup L generated by u is pure. Let $H'' = H' + L$; then H'' is a finite pure subgroup of G and it contains H and x.

Theorem 11. Let G *be a countable primary group with no elements of infinite height. Then* G *is a direct sum of cyclic groups.*

Proof. Number off the elements of G in any way: $x_1, x_2, x_3, \cdots$. By repeated use of Lemma 9 we find a sequence of subgroups H_i with the properties: (*a*) $H_1 \subset H_2 \subset H_3 \subset \cdots$ (*b*) each H_i is a finite pure subgroup of G, (*c*) $x_n \in H_n$. By Theorem 7, H_i is a direct summand of H_{i+1}: say, $H_{i+1} = H_i \oplus K_i$. Then G is a direct sum:

$$G = H_1 \oplus K_1 \oplus K_2 \oplus \cdots \oplus K_n \oplus \cdots .$$

Each summand is finite and is consequently a direct sum of cyclic groups.

It is one of the remarkable features of our subject that Theorem 11 fails if the hypothesis of countability is dropped (see exercise 33). In fact, the classification of uncountable primary groups with no elements of infinite height is a problem on which very little progress has been made. But there is one affirmative result, which we prove next (Theorem 12). It provides a useful characterization of direct sums of cyclic groups, and we shall promptly utilize it to prove another theorem, which has considerable interest of its own. First, a rather technical lemma:

Lemma 10. Let G *be a primary group,* H *a pure subgroup,* x *an element of order* p *not in* H. *Suppose that* $h(x) = r < \infty$, *and suppose further that* $h(x + a) \leq h(x)$ *for every* a *in* H *with* $pa = 0$. *Write* $x = p^r y$, *let*

K *be the cyclic subgroup generated by* y, *and let* L = H + K. *Then* L *is the direct sum of* H *and* K, *and* L *is again pure.*

Proof. That L is the direct sum of H and K merely means that $H \cap K = 0$; and this is immediate from the fact that any nonzero subgroup of K contains x, whereas x is not in H.

To prove that L is pure (in G) it suffices, by Lemma 7, to show that the elements of order p in L have the same height in L and in G. So let w be an element of order p in L. If w is actually in H, then its height is the same in H as in G, since H is pure. Its height in L lies between these two heights and so is the same as either of them.

It remains to consider the case where w is not in H. We can suppose without loss of generality that $w = x + a$ where $a \in H$, $pa = 0$. We know that $h(x) = r$, both in L and in G. Also h(a) has the same meaning, in H, L, or G. If $h(a) \neq r$, then $h(x + a)$ is precisely the minimum of h(a) and r, and this is true in both L and G. Suppose then that $h(a) = r$. Then the height of $x + a$ in L is precisely r. A priori, its height in H might be larger than r, but this would contradict our hypothesis that h(x) is maximal among all $h(x + a)$.

Before going on to the next lemma, we remind the reader that the notation $\Sigma(x_i)$ denotes the subgroup generated by the independent set $\{x_i\}$; thus $\Sigma(x_i)$ is the direct sum of the cyclic subgroups generated by the elements x_i.

Lemma 11. Let G *be a primary group,* P *the subgroup of elements satisfying* $px = 0$. *Let* Q, R *be subgroups of* P, *with* $Q \subset R \subset P$, *and suppose that* R *is of bounded height (i.e., there exists a constant* k *such that* $h(a) \leqq k$ *for all* a *in* R). *Let* $\{x_i\}$ *be a pure independent set satisfying* $\Sigma(x_i) \cap P = Q$. *Then* $\{x_i\}$ *can be enlarged to a pure independent set* $\{y_j\}$ *satisfying* $\Sigma(y_j) \cap P = R$.

Proof. Consider pure independent sets y_j which contain $\{x_i\}$ and satisfy $\Sigma(y_j) \cap P \subset R$. By Zorn's lemma, we may pick a maximal one among these sets; let us again denote it by $\{y_j\}$. We have to prove that $\Sigma(y_j) \cap P$ is actually all of R. Suppose the contrary, and let z be in R but not in $\Sigma(y_j)$. Consider the totality of elements $z + a$, where a ranges over the elements in $\Sigma(y_j)$ satisfying $pa = 0$. Since all elements $z + a$ lie in R, which is of bounded height, we may pick one (say x) of maximal height r. Write $x = p^r y$. Then by Lemma 10 the enlarged set $\{y_j, y\}$ is again pure and independent. Since evidently $\Sigma(y_j, y) \cap P$ is still contained in R, we have contradicted the maximality of y_j.

Our final preliminary lemma belongs in the same family as Lemmas 7 and 8; like them, it is concerned with the dominant role of the elements of order p:

Lemma 12. Let G *be a primary group, and* H *a pure subgroup containing all the elements of order* p *in* G. *Then* H = G.

Proof. Suppose by induction that H is known to contain all elements of order at most p^n, and let x be an element of order p^{n+1}. Then px is in H. By the purity of H, $px = py$ with $y \in H$. By hypothesis $x - y \in H$, and hence $x \in H$.

Theorem 12. Let G *be a primary group,* P *the subgroup of elements satisfying* $px = 0$. *Then a necessary and sufficient condition for* G *to be a direct sum of cyclic groups is that* P *be the union of an ascending sequence of subgroups of bounded height.*

Proof. The necessity is clear. To prove the sufficiency we suppose $P = \cup P_i$, where $P_1 \subset P_2 \subset P_3 \subset \cdots$, and each P_i is of bounded height. We proceed to construct pure independent sets X_i ($i = 1, 2, \cdots$) with the following properties: each X_i is contained in X_{i+1} and, if S_i is the subgroup generated by X_i, then $S_i \cap P = P_i$. If we suppose that $X_1, \cdots, X_r$ have already been constructed, then Lemma 11 applies precisely to enable us to construct X_{r+1}. Now let X be the set-theoretic union of the sets X_i. Then the subgroup generated by X is pure, contains P, and hence (Lemma 12) is all of G. This shows that G is a direct sum of cyclic groups.

Remark. Theorem 12 is clearly more general than Theorems 6 and 11, and in its proof we have not used either of those theorems. Thus it would seem that we could simply suppress the previous discussion of them, but we think that there are valid expository reasons for having presented the (somewhat easier) proofs of Theorems 6 and 11 first.

Theorem 13. Let G *be a primary group which is a direct sum of cyclic groups. Then any subgroup* H *of* G *is a direct sum of cyclic groups.*

Proof. Consider a representation of G as a direct sum of cyclic groups, and let G_i be the union of all cyclic summands of order $\leq p^i$. Let Q be the subgroup of H consisting of elements satisfying $px = 0$. We have $Q = \cup(Q \cap G_i)$; moreover each $Q \cap G_i$ is of bounded height in H (it is even of bounded height in G). It now follows from Theorem 12 that H is a direct sum of cyclic groups.

Remarks. (*a*) Theorem 13 of course generalizes at once to torsion groups. It is even valid for arbitrary groups; this involves some additional considerations (see Theorem 17).

(*b*) The special case of Theorem 13 in which G is countable could be handled by Theorem 11, for it is evident that the property of having no elements of infinite height is inherited by subgroups. The virtue of Theorem 12 in this connection is that it provides a general criterion which is also visibly inherited by subgroups.

Exercises 28-30

28. Let G be a primary group with no elements of infinite height. Show that any finite subgroup of G can be embedded in a finite direct summand.

29. Let G be a primary group with no elements of infinite height.

Show that any subgroup of bounded height in G can be embedded in a direct summand of bounded order.

30. Prove that a countable primary group is a direct sum of indecomposable groups if and only if the elements of infinite height form a pure subgroup.

11. ULM'S THEOREM

The theorem presented in this section is undoubtedly the most striking one yet obtained on abelian groups. It accomplishes nothing less than a complete classification of countable torsion groups. (By Theorem 1 it is enough to do this for primary groups.)

The theorem is remarkable in several respects. It does not say (as one might expect) that any countable primary group is such and such; rather, it says that the group is completely determined by a set of well-defined (though somewhat subtle) invariants. Moreover the very formulation of the theorem involves both cardinal and ordinal numbers in an essential way.

Let us begin by considering what invariants are available. Let G be a primary group. We may then form the subgroups pG, p^2G, $\cdots$, p^nG, $\cdots$ (we have already had occasion to do this in connection with the definition of height). The intersection $\cap p^nG$ is the subgroup of elements of infinite height.

But there is more that we can do. We can take $\cap p^nG$ and multiply it by p, p^2, $\cdots$, thus extending our descending chain of subgroups further. A precise formulation of this procedure is made possible by introducing the transfinite ordinals. (We shall assume that the reader has some familiarity with ordinals, but nothing much is actually needed beyond a knowledge of the definitions and the most elementary properties.)

First, some notation: We write G_n for p^nG ($n = 0, 1, 2, \cdots$), then G_ω for $\cap G_n$, $G_{\omega+1} = pG_\omega$, etc. In general, for any ordinal α we define $G_{\alpha+1}$ to be pG_α, and if α is a limit ordinal we define G_α to be the intersection of G_β for $\beta < \alpha$. There must finally be an ordinal λ such that $G_{\lambda+1} = G_\lambda$ (and from that point on the series remains constant). Since $G_\lambda = pG_\lambda$, G_λ is a divisible subgroup, and it is clear that it is in fact the maximal divisible subgroup of G. Attention may be restricted to reduced groups, and in that case $G_\lambda = 0$. We have built a descending transfinite series beginning at G and ending at 0. We shall refer to the ordinal λ as the *length* of the group G.

In searching for invariants it is natural next to form (for each ordinal $\alpha < \lambda$) the quotient group $G_\alpha/G_{\alpha+1}$. Since $G_{\alpha+1} = pG_\alpha$, this is a group with all its elements of order p, or, if we like, a vector space over the field of integers mod p. Like any vector space, it has a well-defined dimension: a cardinal number which may be finite or infinite.

These cardinals are manifestly invariants of G, but they are too crude to constitute a complete set of invariants. Let us examine their meaning if G is a direct sum of cyclic groups. To do this, we go back even further, to the case where G is itself cyclic. Then if G has order $\leq p^n$, we have $G_n = p^n G = 0$. If G has order $> p^n$, then G_n/G_{n+1} is clearly cyclic of order p. Now since the computation of G_n/G_{n+1} goes coördinatewise, we see that we get one cyclic group of order p for each cyclic summand of order $> p^n$. In summary: If G is a direct sum of cyclic groups, then the dimension of G_n/G_{n+1} *(as a vector space over the integers mod* p) *is the number of cyclic summands of order* $\geq p^{n+1}$.

Now if G is a finite group, these dimensions are a complete set of invariants, for, by subtracting successive pairs, we get precisely the number of cyclic summands of each order. However, such subtraction is no longer feasible in the infinite case. Let us give an explicit example. If G is the direct sum of cyclic groups of order p, p^2, p^3, $\cdots$, then each G_n/G_{n+1} is infinite and has dimension $\aleph_0$. If, on the other hand, G is the direct sum of two copies each of cyclic groups of order p, p^2, p^3, $\cdots$, then again G_n/G_{n+1} has dimension $\aleph_0$ for all n. Hence these two groups cannot be distinguished by the invariants given by the dimensions of G_n/G_{n+1}.

The clue to an improved set of invariants lies in an observation, made several times earlier, that the elements of order p play a special role in determining the behavior of the entire group. We accordingly introduce the group P of all elements satisfying $px = 0$, and we set $P_\alpha = P \cap G_\alpha$; more generally, if S is *any* subgroup of G, we define $S_\alpha = S \cap G_\alpha$. The quotient group $P_\alpha/P_{\alpha+1}$ may be regarded as a vector space over the integers mod p, and we write $f(\alpha)$ for its dimension; if it becomes necessary to call attention to the dependence upon G, we shall write $f(\alpha, G)$. We call $f(\alpha)$ the α-th *Ulm invariant* of G. Thus the Ulm invariants constitute a function from the ordinals to the cardinals.

Let us consider again the case where G is a cyclic group, say of order p^k. If $k \leq n$, then G_n, P_n, and P_n/P_{n+1} are all 0, so that $f(n) = 0$. If $k \geq n + 2$, then G_n is a cyclic group of order p^{k-n}, G_{n+1} is its unique subgroup of order $p^{k-n-1} \geq p$, and so both P_n and P_{n+1} are cyclic of order p, $P_n/P_{n+1} = 0$, and $f(n) = 0$. Only when k is precisely $n + 1$ do we get anything nonzero, and we find $f(k - 1)$ to be precisely 1. Adding up the components, we arrive at the following conclusion: *If* G *is a direct sum of cyclic groups, then its* n*-th Ulm invariant* f(n) *is precisely the number of cyclic summands of order* p^{n+1}. Thus, at least for direct sums of cyclic groups, the Ulm invariants do provide a complete characterization. Incidentally, we have at length proved that the expression of a primary group as a direct sum of cyclic groups (whenever one exists) is unique up to isomorphism.

We are now ready to state Ulm's theorem:

Theorem 14. Two reduced countable primary abelian groups are isomorphic if and only if they have the same Ulm invariants.

Before giving the proof it will be useful to make several observations. It is to be understood that G is a reduced primary group, but the hypothesis of countability will not be needed for a while.

First, we must refine our previous definition of height so as to make some distinction between the elements of infinite height. For $x \neq 0$ we define the height of x, written h(x), to be α if x is in G_α but not in $G_{\alpha+1}$. This assigns to each nonzero element a well-defined ordinal less than λ, the length of G. Note that for elements of finite height, our new definition coincides with the old. As for the element 0, it turns out to be desirable to write $h(0) = \infty$, with the convention that ∞ exceeds any ordinal. (At any rate it is unwise to write $h(0) = \lambda$; $h(0) = \lambda + 1$ would be acceptable.)

It is immediately clear that our fundamental inequalities concerning height survive in this refined context. That is:

(*a*) If $h(x) < h(y)$, then $h(x + y) = h(x)$.
(*b*) If $h(x) = h(y)$, then $h(x + y) \geqq h(x)$.

We may also assert:

(*c*) If $x \neq 0$, then $h(px) > h(x)$.

If S is any subgroup of G and x is any element in G, we shall say that x is *proper* with respect to S if $h(x) \geqq h(x + s)$ for every $s \in S$, or, in other words, if x has maximal height in its coset mod S. Note that under these circumstances $h(x + s)$ is determined by h(x) and h(s), being, indeed, the minimum of the two. Note further that if S is finite, there will certainly be a proper element in each coset.

In preparation for stating a crucial lemma, we now set up a certain isomorphism. Let S be any subgroup of G, and α any ordinal. Write $S_\alpha{}^* = S_\alpha \cap p^{-1}G_{\alpha+2}$. (We recall that S_α is our notation for $S \cap G_\alpha$; $p^{-1}G_{\alpha+2}$, of course, means the set of all z such that $pz \in G_{\alpha+2}$.) The idea is roughly this: For any x in S_α we know that px is in $S_{\alpha+1}$. Usually this is the most one can say, but there will be some elements that are carried at least one step further down, and these we are calling $S_\alpha{}^*$. We note that for any x in $S_\alpha{}^*$, px may be written py, with $y \in G_{\alpha+1}$. The element y is not unique; indeed, it is permissible to change it by any element of order p in $G_{\alpha+1}$, that is, by any element in $P_{\alpha+1}$. It is more convenient for us to work with x - y; since px = py, $x \in G_\alpha$, and $y \in G_{\alpha+1}$, we have $x - y \in P_\alpha$. Also x - y, like y, is well-defined modulo $P_{\alpha+1}$. In sum, the mapping $x \to x - y$, followed by the natural homomorphism from P_α to $P_\alpha/P_{\alpha+1}$, is a homomorphism of $S_\alpha{}^*$ into $P_\alpha/P_{\alpha+1}$. The kernel is precisely $S_{\alpha+1}$. We now have an isomorphism, which we shall call U, of $S_\alpha{}^*/S_{\alpha+1}$ into $P_\alpha/P_{\alpha+1}$.

The following lemma will doubtless seem somewhat artificial; however, it is just what is needed in the proof of Theorem 14:

Lemma 13. If U *is the mapping just defined, then the following two statements are equivalent:*

(*a*) *The range of* U *is not all of* $P_\alpha/P_{\alpha+1}$.

(*b*) *There exists in* P_α *an element of height* α *proper with respect to* S.

Proof. (*a*) → (*b*). Let v be an element of P_α whose coset (mod $P_{\alpha+1}$) is not in the range of U. Then, since v is surely not in $P_{\alpha+1}$, its height is precisely α. We shall show that v is proper with respect to S. Suppose the contrary. Then there must exist an element s in S with $h(s - v) > \alpha$. We may write $s - v = pt$ with t in G_α. Since $pv = 0$, we have $ps = p^2t \in G_{\alpha+2}$. It follows that s is in $S_\alpha{}^*$. Now if we consult the definition of the mapping U, we find that we begin by sending s into $s - pt = v$, and then take cosets. In other words, U sends the coset of s (mod $S_{\alpha+1}$) into the coset of v (mod $P_{\alpha+1}$). This is a contradiction, since the coset of v was assumed not to be in the range of U.

(*b*) → (*a*). Suppose v is in P_α, has height α, and is proper with respect to S. Then the coset of v is not in the range of U. For if it were, there would have to exist elements x in S, y in $G_{\alpha+1}$ with $p(x - y) = 0$ and

$$v \equiv x - y \qquad\qquad (\text{mod } P_{\alpha+1}).$$

Then $h(v - x) > \alpha$, in contradiction to the assumption that v is proper with respect to S.

Proof of Theorem 14. Let G and H be countable reduced primary groups with the same Ulm invariants. Our task is to prove G and H isomorphic.

Before we finally give the proof, several further remarks may be helpful:

(*a*) In order to make use of the hypothesis of countability, the isomorphism of G and H will be constructed step by step. At every stage of the argument we shall then have to worry only about finite subgroups of G and H. Without countability, it would be impossible to get through the proof in this way. Instead we should have to use a transfinite process, and at the intermediate stages we would encounter infinite groups. Since Ulm's theorem is false without the countability hypothesis, it is predictable that such a transfinite method must fail.

(*b*) Among other things, we have to make certain that our construction yields an isomorphism between all of G and all of H. To do this, we number off, once for all, the elements of G and H. Then at the (2n - 1)-th step, we look after the n-th element of G; at the 2n-th step, we look after the n-th element of H. This alternation between G and H is an indispensable device. If, for instance, we confined our attention to G, it is perfectly possible that we would wind up with an isomorphism between G and *part* of H.

(*c*) The completed isomorphism between G and H must of course be height-preserving. This indicates the advisability of making certain, as we proceed, that heights are being preserved. If we were so careless as to begin with a mapping that failed to preserve heights (and it is quite possible to do so), the desired isomorphism could never be completed. (It is to be understood that heights are always computed relative to G and H, never relative to intermediate subgroups.)

In the light of these remarks we may assume the following situation: S and T are finite subgroups of G and H, respectively, and V is a height-preserving isomorphism of S onto T; S' is the subgroup generated by S and an element x in G. Our aim is to extend V to a height-preserving isomorphism of S' and a suitable subgroup of H containing T. We may suppose that x is not in S, but that px is in S.

We now make two further normalizing assumptions: We assume that x is proper with respect to S, i.e., that $\alpha = h(x)$ is maximal in $x + S$; then we assume that among all elements in $x + S$ for which this is true, h(px) is maximal. These normalizations cause no trouble, since there are only a finite number of such elements.

Write $(px)V = z$. Our problem comes down to this: Find in H an element w which satisfies $pw = z$, has height α, and is proper with respect to T. For once we have found w, we simply extend V by sending x into w—this will still be an isomorphism and height-preserving. We distinguish two cases:

Case I. $h(z) = \alpha + 1$. It is to be noted that by our convention about the height of 0, z (and hence px) must be nonzero. We now choose for w any element in H_α with $pw = z$. First, h(w) cannot exceed α, for, if it did, $h(z) = h(pw)$ would exceed $\alpha + 1$. Next, it cannot be true that w is in T. For suppose $w = yV$, with y in S. Then $px = py$, since both map on z; also $x - y$ is not in S, for otherwise x would be in S; and $h(x - y) = \alpha$, since $h(y) = \alpha$ and x is proper with respect to S. But $h(px - py) = h(0) > \alpha + 1$, and this contradicts the maximal choice of h(px). Thus w is not in T It remains for us to prove that w is proper with respect to T. Suppose, on the contrary, that $h(w + t) \geqq \alpha + 1$ where $t \in T$, $t = sV$ for $s \in S$. Since $w + t \neq 0$, we have $h(pw + pt) \geqq \alpha + 2$, hence $h(px + ps) \geqq \alpha + 2$. Now t necessarily has height at least α, hence so has s, and $h(x + s) = \alpha$. We have again contradicted the maximality of h(px).

(Case I is in some sense the easier of the two. Note that to dispose of it, we did not need the equality of the Ulm invariants.)

Case II. $h(z) > \alpha + 1$. Now $h(px) > \alpha + 1$ means that $px = pv$ with $v \in G_{\alpha+1}$. The element $x - v$ is then in P_α; like x, it has height α, and so it is likewise proper with respect to S (since v does not interfere in computations of heights $\leqq \alpha$). We can now apply Lemma 13. Since $S_\alpha{}^*/S_{\alpha+1}$ is finite, Lemma 13 tells us that its dimension (as a vector space over the integers mod p) is strictly less than the Ulm invariant $f(\alpha)$. Since V is height-preserving, it maps (in the obvious notation) S_α onto T_α, $S_\alpha{}^*$ onto $T_\alpha{}^*$, and $S_\alpha{}^*/S_{\alpha+1}$ onto $T_\alpha{}^*/T_{\alpha+1}$. So the dimension of $T_\alpha{}^*/T_{\alpha+1}$ is likewise less than $f(\alpha)$—which by hypothesis is also the α-th Ulm invariant of H. We apply Lemma 13 again and deduce that H contains an element w_1 which satisfies $pw_1 = 0$, has height α, and is proper with respect to T. Next we note that since $h(z) > \alpha + 1$, z may be written pw_2 with $w_2 \in H_{\alpha+1}$. We are ready to take $w = w_1 + w_2$. It has the desired properties: $pw = z$, $h(w) = \alpha$, and w is proper with respect to T. This concludes the proof of Theorem 14.

We shall close this section with several remarks:

(*a*) Ulm's theorem is amply strong to settle both of our test problems for countable torsion groups—see exercise 32.

(*b*) Our discussion of Ulm's theorem has been notably poor in examples. Thus it is appropriate to give at least one, an example which, in a sense, is the simplest possible. It seems inevitable that we can do so only by specifying generators and relations.

We take elements x, y_i (i = 1, ⋯ , n, ⋯) and subject them to the relations: $px = 0$,

$$x = py_1 = p^2y_2 = \cdots = p^ny_n = \cdots .$$

The Ulm invariants are

$$f(0) = f(1) = f(2) = \cdots = f(n) = \cdots = f(\omega) = 1,$$

and the group has length $\omega + 1$.

(*c*) The argument of Theorem 14 actually proved more; in particular, it proved a result on extensions of isomorphisms which we shall actually use later. In order not to interrupt the exposition, we have inserted this result as exercise 38, below. A corresponding assertion concerning the extension of homomorphisms turns out to be easier to prove; we leave it to the reader as exercise 39.

(*d*) The question of the existence of groups with prescribed Ulm invariants deserves investigation on almost the same basis as that of uniqueness. See on this the information provided in the exercises. In the first place (exercise 36), it is easy to prove the following necessary condition: Between any two limit ordinals (less than the length of the group) there must exist an infinite number of nonvanishing Ulm invariants. For countable groups, it turns out that this is the sole condition that has to be imposed (exercise 42), and thus the theory of countable primary groups reaches a satisfactory conclusion. In the uncountable case not much is known, but in exercise 44 we indicate that further conditions have to be imposed on sets of cardinals before they are eligible to be Ulm invariants.

Exercises 31-45

31. Let G and H be reduced primary groups. Show that the Ulm invariants of G ⊕ H are obtained by adding the Ulm invariants of G to those of H; that is, $f_\alpha(G \oplus H) = f_\alpha(G) + f_\alpha(H)$. Generalize to any number of summands.

32. (*a*) In either of the two test problems (§6), let G and H be reduced primary groups. Show that we can at any rate conclude that G and H have the same Ulm invariants. (Use exercise 31.)

(*b*) Assuming further that G and H are countable, deduce that G and H are isomorphic.

(*c*) Answer both problems in the affirmative for countable torsion groups.

33. Let G be the *complete* direct sum of cyclic groups of orders p, p^2, p^3, $\cdots$, and let T be its torsion subgroup.

(*a*) Show that T is not a direct summand of G. (Observe that G has no elements of infinite height, but that G/T does.)

(*b*) Show that T is a primary group with no elements of infinite height, but that it is not a direct sum of cyclic groups. (Prove that the Ulm invariants are 1, 1, 1, $\cdots$ and deduce that T would be countable if it were a direct sum of cyclic groups. Note that this is an example showing that countability cannot be dropped in Theorems 11 and 14. As a counterexample to Theorem 14 it is open to the objection that at the very least we ought to be comparing two groups with the same cardinal number. But this objection is easily answered by taking the direct sum with a group whose cardinal is large enough to swallow up the difference.)

34. Let G be a direct sum of cyclic groups. Prove that any two direct decompositions of G have isomorphic refinements. (Use Theorem 13.)

35. Let G be a reduced primary group of length λ, and let α and β be ordinals with β a limit ordinal and $\alpha < \beta \leq \lambda$. Show that the Ulm invariants of G_α/G_β coincide with the Ulm invariants of G between α and β. Is this true if β is not a limit ordinal?

36. Let G be a reduced primary group of length λ. Let α and β be limit ordinals with $\alpha < \beta \leq \lambda$. Show that the Ulm invariants of G must be nonzero for an infinite number of ordinals between α and β. (Use exercise 35.)

37. Let G be a reduced primary group. Consider the invariants given by the dimensions of $G_\alpha/G_{\alpha+1}$ ($\alpha<\omega$). Show that the analogous invariants for any homomorphic image of G are smaller. Show that the same is true for a subgroup of G if G is finite.

38. Let G be a countable reduced primary group, S and T finite subgroups. Let V be a height-preserving isomorphism of S and T. Show that V can be extended to an automorphism of G. (This was actually proved in the course of proving Theorem 14.)

39. Let G and H be reduced primary groups with G countable, and let S be a finite subgroup of G. Let V be a homomorphism of S into H which is height-increasing in the sense that $h(sV) \geq h(s)$ for all s in S. Show that V can be extended to a height-increasing homomorphism of all of G into H. (This is easier than exercise 38. We have to extend V to an element x, where we can assume that $px \in S$ and that x is proper with respect to S. We need only take xV to satisfy $p(xV) = (px)V$ and $h(xV) \geq h(x)$, and this is readily seen to be possible.)

40. Let λ_i be a monotone increasing sequence of ordinals ap-

proaching λ. Let n be a given nonnegative integer. For each i let there be given a primary group G(i) such that $G(i)_{\lambda_i}$ is cyclic of order p^{n+2}, generated, say, by x_i. (Note this implies that the $(\lambda_i + n + 1)$-th Ulm invariant of G(i) is 1.) Let H be the direct sum of the groups G(i), reduced by identifying the elements px_i. What are the Ulm invariants of H? (Answer: Add the Ulm invariants of the G's, deleting the $(\lambda_i + n + 1)$-th invariant of G(i); insert a $(\lambda + n)$-th invariant equal to 1 and, for each i, a λ_i-th invariant equal to 1.)

41. Let f be a function for ordinals to cardinals, defined up to the limit ordinal λ. Call f *admissible* if it takes on an infinite number of nonzero values between any two limit ordinals $\leq \lambda$. (This definition of admissibility is suggested by exercise 36.) Let λ_i be a monotone increasing sequence of ordinals approaching λ. Show that f can be written as the sum of a countable number of admissible functions f_i, where f_i vanishes after λ_i.

42. Let f be an admissible function, defined up to a countable ordinal λ, and taking values $\leq \aleph_0$. Prove that there exists a countable primary group with Ulm invariants given by f. (Argue by induction on λ. When λ is a limit ordinal, use exercise 41 and take the direct sum of the appropriate groups whose existence is known by introduction. When λ is not a limit ordinal, take the direct sum of a finite number of groups constructed with the aid of exercise 40.)

43. Show that for any ordinal λ there exists a reduced primary group of length λ. (Use the technique of exercise 40.)

44. Let G be a reduced primary group such that for each finite i its Ulm invariant satisfies $f(i) \leq \aleph_0$. Show that $f(\omega) \leq \aleph_0$, also. (This indicates that more conditions must be imposed on proposed Ulm invariants than the one in exercise 36.)

45. Show that there exists a reduced primary group of length Ω (the first uncountable ordinal) with each Ulm invariant equal to $\aleph_1$ (the first uncountable cardinal). (Take the direct sum of $\aleph_1$ suitable countable groups.)

12. MODULES AND LINEAR TRANSFORMATIONS

The purpose of this section is to indicate how all the preceding results extend to modules over principal ideal rings and to apply this to the theory of linear transformations.

Let R be an integral domain (commutative ring with unit and no divisors of zero). An abelian group M is said to be an R-*module* if there is a product λx defined for $\lambda \in R$, $x \in M$ satisfying

$$\begin{aligned} \lambda(x + y) &= \lambda x + \lambda y, \\ (\lambda + \mu) x &= \lambda x + \mu x, \\ (\lambda \mu) x &= \lambda(\mu x), \\ 1 x &= x. \end{aligned}$$

A subset S is said to be a *submodule* if S is, again, an R-module relative to the same operations; or, in other words, S is an additive subgroup satisfying $\lambda S \subset S$ for every λ in R. For any submodule S of M we can define a *quotient module* M/S: additively it is simply the quotient group, and we define the product of the coset x + S by $\lambda \in R$ to be $\lambda x + S$ (this definition being independent of the choice of the representative x). An R-module M is said to be *cyclic* if there exists a single element generating M, i.e., if M = Rx for some x in M.

If M and M' are R-modules, a homomorphism (one might speak of an R-homomorphism to avoid any ambiguity) is an additive mapping $x \to x'$ of M into M' which satisfies $(\lambda x)' = \lambda x'$ for all λ in R. If the kernel of the homomorphism is K and the image N, then K is a submodule of M and M/K is isomorphic to N.

For any x in an R-module M, the set of all $\lambda \in R$ with $\lambda x = 0$ forms an ideal in R, called the *order ideal* of R. The set of all elements in M with a nonzero order ideal forms a submodule T of M called the *torsion submodule*. If T = M, we say that M is a *torsion module*; if T = 0, we say that M is *torsion-free*. For any arbitrary M, the quotient module M/T is torsion-free.

Let us pause to give some examples:

(*a*) Let R be the ring of integers, M any abelian group. For n an integer and $x \in M$, let nx have its usual meaning. Then M is an R-module. All the concepts defined above coincide in an evident fashion with the corresponding ones for plain abelian groups. In this way, the theory of abelian groups appears as a special case of the theory of modules.

(*b*) R itself is an R-module, if we use the multiplication already present in R. The submodules are precisely the ideals in R. For any ideals J and I with $J \subset I$, we can further regard I/J as an R-module.

Let us observe that any cyclic R-module M is isomorphic to an ideal of the form R/I. For let M = Rx, and let I be the order ideal of x. The mapping $\lambda \to \lambda x$ (for $\lambda \in R$) is a homomorphism of R onto M, with kernel I. Hence M is isomorphic to R/I.

(*c*) Suppose R is a field. Then an R-module is exactly the same thing as a vector space over R. Any vector space (finite- or infinite-dimensional) has a basis; in other words, a vector space over R is a direct sum of one-dimensional subspaces, these latter being submodules isomorphic to R. The structure problem for R-modules is thus trivial if R is a field.

(*d*) We come now to the example which is the real justification for our having introduced modules at all. Let V be a vector space (any dimension) over a field F, and let T be a linear transformation of V; we shall write T on the left of vectors in V. Let R = F[u] be the ring of polynomials in a variable u, with coefficients in F. We construe V to be an R-module as follows: For $x \in V$ and $f(u) \in R$ we define f(u)x to be f(T)x. In greater detail, if

$$f(u) = \alpha_0 + \alpha_1 u + \cdots + \alpha_m u^m,$$

then

$$f(u)x = f(T)x = \alpha_0 x + \alpha_1(Tx) + \cdots + \alpha_m(T^m x).$$

It is to be noted that R contains F, and when R acts on V, the elements of F act as scalars on the vector space. In other words, we have supplemented the action of F on V by inserting the action of T on V.

It is routine to verify that the definition just given does indeed make V an R-module. When desirable, we shall denote this module by V_T, thus indicating the dependence on T.

Let us see what it means for V_1 to be a submodule of V_T. Since R contains F, V_1 must, in the first place, be a subspace; further, V_1 must be closed under the action of u, that is to say, T. In other words, V_1 must be an invariant subspace of V.

If V_T is a module direct sum $V_1 \oplus V_2$, then we have a direct-sum decomposition of V into two invariant subspaces.

If a basis is chosen for the vector space V, we get a matrix representing T relative to this basis. If further, as above, V_1 is a submodule (= invariant subspace) of V_T, and if we choose a basis for V which contains a basis of V_1, then the matrix of T has the following form:

$$\begin{pmatrix} * & 0 \\ * & * \end{pmatrix}$$

If $V_T = V_1 \oplus V_2$, and we complete a basis of V by using a basis of V_2, there will be zeros, as well, in the lower left corner. (If V is infinite-dimensional, we are speaking of infinite matrices here, in a sense that need not be made too precise.)

Suppose now we are given a second linear transformation S on the same vector space V. As is customary, we say that S and T are *similar* if there exists a nonsingular linear transformation P (i.e., one with a two-sided inverse) such that $P^{-1}SP = T$. *Then* V_S *and* V_T *are isomorphic* R*-modules if and only if* S *and* T *are similar.* The proof of this merely requires examination of the relevant definitions. Suppose P is an isomorphism of V_S onto V_T. Then P is an additive mapping of V onto itself, which preserves multiplication by elements of R. But multiplication by $F(u) \in R$ means applying f(S) or f(T), according as we are looking at V_S or V_T. Hence

$$P[f(S)x] = f(T)(Px) \tag{7}$$

for every x in V. If we apply equation (7) with f a scalar (i.e., an element of F), we deduce that P is a linear transformation. Next take the case where $f(u) = u$. We obtain $PS = TP$, $PSP^{-1} = T$. The argument may be reversed so as to show that similarity of S and T implies isomorphism of V_S and V_T.

The problem of classifying linear transformations under similarity is thus seen to be a special case of the classification of R-modules under isomorphism, namely, the special case where R is of the form F[u].

Matters now stand as follows: We would like to discuss modules over two rings, the ring of integers and the ring of polynomials in one variable over a field. The appropriate concept for unifying these two concepts is that of a principal ideal ring (integral domain in which every ideal is principal). We may summarize the known facts which we need (they appear in the standard treatises on modern algebra) as follows:

(*a*) In a principal ideal ring every set has a greatest common divisor, unique up to a unit, and it is expressible as a linear combination of the elements.

(*b*) Every element (not 0 or a unit) is expressible as a product of prime elements, and the expression is unique, up to order and insertion of units.

(*c*) If p is a prime element, the ring R/(p) is a field.

(*d*) Any integral domain with a Euclidean algorithm is a principal ideal ring. (See [95] for an example of a principal ideal ring in which there can be no Euclidean algorithm, even in a certain weak sense.)

We are ready to make a blanket assertion: *Theorems 1—14 hold for modules over principal ideal rings.*

Ideally, we should have developed the entire subject, right from the start, for modules over an arbitrary principal ideal ring; it was only for expository reasons that the introduction of modules was delayed till now. At any rate, we shall not repeat any proofs, for only nominal changes are needed. Among these nominal changes, two are perhaps worth singling out for special attention.

First, the words "finite" and "countable" need translation: they become "finitely generated torsion" and "countably generated." (It is understood that unless R itself is countable, we cannot expect an R-module to be actually countable.) In the linear-transformation case the translations read simply "finite-dimensional" and "of countable dimension."

Second, there is one point in the proof of Theorem 14 which occasions a little trouble. We have a finitely generated primary submodule S and are engaged in adjoining an element x to it; we have to normalize by choosing x to have maximal height in the coset x + S. Now that S is not actually finite, but merely finitely generated, it is not so clear that such a maximal element exists. However, it is a fact that there are still really only a finite number of elements from which to choose; we leave this to the reader to verify.

In the remainder of this section we shall consider in greater detail the interpretation of the theory in the case of a linear transformation. We begin with a "dictionary" and some explanations concerning it:

Group	*Linear transformation*
torsion	locally algebraic
bounded order	algebraic
primary	$\lambda I + U$, U locally nilpotent
primary, bounded order	$\lambda I + U$, U nilpotent
direct sum of finite cyclic groups	direct sum of finite-dimensional invariant subspaces
infinite cyclic	$\begin{pmatrix} 0 & 1 & 0 & 0 & \cdots \\ 0 & 0 & 1 & 0 & \cdots \\ 0 & 0 & 0 & 1 & \cdots \\ & \cdot & \cdot & \cdot & \\ & \cdot & \cdot & \cdot & \\ & \cdot & \cdot & \cdot & \end{pmatrix}$
cyclic, finite order	$\begin{pmatrix} 0 & 1 & 0 & 0 & \cdots & 0 \\ 0 & 0 & 1 & 0 & \cdots & 0 \\ 0 & 0 & 0 & 1 & \cdots & 0 \\ \cdot & \cdot & \cdot & \cdot & & \\ 0 & 0 & 0 & 0 & & 1 \\ \alpha_n & \alpha_{n-1} & & & \cdots & \alpha_1 \end{pmatrix}$
cyclic, prime power order	$\begin{pmatrix} \lambda & 1 & 0 & \cdots & & 0 \\ 0 & \lambda & 1 & \cdots & & 0 \\ 0 & 0 & \lambda & \cdots & & 0 \\ \cdot & \cdot & \cdot & & & \\ 0 & 0 & 0 & & \lambda & 1 \\ 0 & 0 & 0 & \cdots & & \lambda \end{pmatrix}$
$Z(p^\infty)$	$\begin{pmatrix} \lambda & 0 & 0 & 0 & \cdots \\ 1 & \lambda & 0 & 0 & \cdots \\ 0 & 1 & \lambda & 0 & \cdots \\ 0 & 0 & 1 & \lambda & \cdots \\ \cdot & \cdot & \cdot & & \\ \cdot & \cdot & \cdot & & \end{pmatrix}$

(*a*) To say that V_T is a torsion module is equivalent to saying: For any x in V there exists a nonzero polynomial f such that $f(T)x = 0$. In line with customary modern terminology, we say that T is *locally algebraic.*

(*b*) To say that V_T is of bounded order is to assert the existence of a fixed polynomial f with f(T) = 0. We say that T is *algebraic.*

(*c*) Before looking at the primary case, we recall that the primes in the ring F[u] are precisely the irreducible polynomials. Now a highly desirable simplification occurs if we assume that F is algebraically closed, for then the irreducible polynomials are linear. We shall make that assumption from now on. Then the assertion that V_T is a primary module means that there is a scalar λ such that every x is annihilated by some power of $T - \lambda I$, where I is the identity linear transformation. Or, if we write $T = \lambda I + U$, we have that U is *locally nilpotent* in the sense that every vector is annihilated by some power of U.

(*d*) Suppose V_T is a cyclic module, generated by the vector x. In the infinite cyclic case the vectors $\{T^i x\}$ are linearly independent and form a basis for V. If V_T has the nonzero order ideal (f), where

$$f(u) = u^n - \alpha_1 u^{n-1} - \alpha_2 u^{n-2} - \cdots - \alpha_n,$$

we choose the basis x, Tx, $\cdots$, $T^{n-1}x$ for V; in terms of this basis, T is represented by the so-called companion matrix shown above.

We leave to the reader the discussion of the primary cyclic case, and of $Z(p^\infty)$. But we shall explicitly state the analogues of four of our earlier theorems:

Theorem 1'. If T *is locally algebraic on* V, *then* V *is a direct sum of spaces* $V(\lambda)$, *and on* $V(\lambda)$, $T - \lambda I$ *is locally nilpotent.*

Theorem 6'. If T *is an algebraic linear transformation on* V, *then* V *is a direct sum of finite-dimensional invariant subspaces.*

Theorem 11'. If T *is a locally nilpotent linear transformation on a vector space* V *of countable dimension, and if the intersection of the ranges of the powers of* T *is* 0, *then* V *is a direct sum of finite-dimensional invariant subspaces.*

Theorem 13'. If V *is a direct sum of finite-dimensional subspaces under* T, *then any invariant subspace of* V *is likewise a direct sum of finite-dimensional invariant subspaces.*

In conclusion, we mention the Ulm invariants. The appropriate context is a locally nilpotent linear transformation T on a vector space V, with the added assumption that T is reduced, i.e., there exists no invariant subspace V_1 with $T(V_1) = V_1$. Then we form, as before, the Ulm invariants: a function from ordinal numbers to cardinal numbers. If V is a direct sum of finite-dimensional invariant subspaces, the Ulm invariants amount to the same thing as the elementary divisors (or, more precisely, the elementary divisors for the characteristic root zero, the latter being the only characteristic root of a locally nilpotent linear transformation). If V is of countable dimension, the Ulm invariants are a *complete* set of invariants. And even if the dimension is not countable, they are at any rate invariants. One might say that the Ulm invariants are the last word on the subject of elementary divisors.

A word of explanation may be helpful concerning the other set of invariants which is widely used in the finite-dimensional case, the so-called *invariant factors.* For the sake of simplicity we may as well revert to ordinary abelian groups. Then we know that any finite abelian group is a direct sum of cyclic groups, and that the decomposition is unique when broken down to prime-power order. But there is another way of making the decomposition unique, namely, by arranging that the order of each summand divide the order of its successor. For example, we have the two arrangements:

$$Z_2 \oplus Z_6 \oplus Z_{12} = Z_2 \oplus Z_2 \oplus Z_4 \oplus Z_3 \oplus Z_3.$$

One calls 2, 6, 12 the invariant factors, while 2, 2, 4, 3, 3 are the elementary divisors.

It is a matter of taste which arrangement is used in the finite case. But in the infinite case we have no choice. Outside of the sheer impossibility of bringing about the invariant-factor arrangement in general, we do not even get uniqueness—observe that the groups

$$Z_2 \oplus Z_6 \oplus Z_6 \oplus \cdots$$

and

$$Z_6 \oplus Z_6 \oplus Z_6 \oplus \cdots$$

are isomorphic.

In sum: The decomposition into primary parts, while optional in the finite case, is indispensable in the infinite case.

Exercises 46-48

46. Let M be a module over a principal ideal ring R. Define the *spectrum* to be the set of all primes p in R such that $x \to px$ is not an automorphism of M.

(*a*) If M is a torsion module, show that the spectrum consists of those primes for which there is a nonzero primary summand.

(*b*) Suppose M is not a torsion module, and can be generated by $\aleph$ elements. Show that the set of primes not in the spectrum has cardinal at most $\aleph$. (Take a maximal set of linearly independent elements generating a free submodule P. Prove that any p which does not occur in the torsion module M/P cannot induce an automorphism of M.)

47. Let T be a linear transformation on an $\aleph$-dimensional vector space over an algebraically closed field. Define the spectrum to be the set of all λ such that $(T - \lambda I)^{-1}$ does not have a two-sided inverse, and observe the identity of this concept with that in exercise 46.

(*a*) If T is locally algebraic, show that the spectrum has at most cardinal $\aleph$.

(*b*) If T is not locally algebraic, show that the complement of the spectrum has cardinal at most $\aleph$.

(These results were proved by Ulm [120] by his infinite-matrix methods. He confined himself to vector spaces of countable dimension over the complex numbers, and so his version has a more special form.)

48. (*a*) Let T be a linear transformation on a vector space over a field of characteristic $\neq 2$, and suppose $T - I$ is locally nilpotent. Show that T and T^2 have the same Ulm invariants. (Use the actual definition of the Ulm invariants. Note that $T^2 - I = (T - I)(T + I)$, where $T + I$ is nonsingular, that is, has a two-sided inverse.)

(*b*) Assuming further that V has countable dimension, prove that T is similar to T^2. (After the maximal divisible subspace is disposed of, part (*b*) is an immediate consequence of part (*a*). Whether part (*b*) is valid without countability makes an interesting test problem.)

(*c*) Let T be a nonsingular locally algebraic linear transformation on a vector space of countable dimension over an algebraically closed field of characteristic $\neq 2$. Prove that T has a square root. (Reduce to the primary case, adjust the characteristic root to be 1, and cite part (*b*).)

(We have inserted (*c*) as an application of (*b*). But (*c*) admits a direct proof: if T is locally nilpotent, we can make sense out of the power series for $\sqrt{I + T}$.)

13. BANACH SPACES

In the last section we obtained (incidentally to considering modules) a number of results concerning linear transformations on infinite-dimensional vector spaces. Now there exists, as well, a rather extensive theory of continuous linear transformations (= bounded operators) on Banach spaces. We shall devote this brief section to an attempt to explain why there is virtually no connection between these two theories.

In the first place, we confined ourselves entirely to locally algebraic linear transformations. On a Banach space, it turns out that such a linear transformation must be actually algebraic:

Theorem 15. A continuous locally algebraic linear transformation T *on a Banach space* V *is actually algebraic.*

Proof. The proof is a typical category argument. Let S_n denote the set of all x in V for which there exists a polynomial f of degree $\leq n$ with $f(T)x = 0$. We claim that S_n is closed. Suppose that $x_i \in S_n$, $x_i \to x$, $f_i(T)x_i = 0$, where the degree of f_i is at most n. We can normalize so that all coefficients of f_i are at most 1 in absolute value, and one coefficient is precisely 1. Then a suitable subsequence of the f's will approach a nontrivial polynomial g, and we have $g(T)x = 0$. Hence, S_n is closed.

Since V is the union of the sets S_n, the Baire category theorem tells us that some S_n has a nonvoid interior U. Choose any y in U. Then U - y is a neighborhood of 0, and any of its elements is annihilated by a polynomial of degree at most 2n. Since we can expand this neighborhood by multiplying by scalars, the same will be true for every vector in V. The rest of our task is purely algebraic, and we formulate it as a separate lemma:

Lemma 14. Let T *be a linear transformation on a vector space* V, *and suppose that for any* x *in* V *there exists a polynomial* f *of degree at most* m *such that* $f(T)x = 0$. *Then* T *is algebraic.*

Proof. Split V into primary components by Theorem 1. There can be at most m of these components. This reduces the problem to the primary case, where the proof is evident.

(In the proof of Theorem 15 we have made only meager use of the properties of the field of scalars (real or complex numbers). Actually, the theorem is valid over any complete field with a valuation.)

So we now consider an algebraic linear transformation T on a Banach space. We perform the decomposition into primary subspaces of Theorem 1; there are only a finite number of summands, each of them closed. Thus this part of the theory is satisfactory in a Banach space V—although if V is a Hilbert space, one would prefer a decomposition into orthogonal subspaces; such a decomposition will occur in general only if T is normal.

We have reached the situation where T is nilpotent. In the purely algebraic case, a certain finite set of cardinal numbers (the multiplicities of the elementary divisors, or, if we prefer, the Ulm invariants) are a complete set of invariants. This is not so if V is a Banach space. One might explain this in two ways:

(*a*) The algebraic theory gives us a decomposition of V into finite-dimensional invariant subspaces. But this decomposition is one where only finite sums are allowed. Thus in the algebraic case we are paying no attention to the infinite sums which are possible in a Banach space.

(*b*) The algebraic theory solves the problem of similarity under *any* nonsingular linear transformation. In a Banach space we must stick to similarity under *continuous* linear transformations. This (apparently slight) change in the problem actually makes an enormous difference. We have passed from an easy question to what appears to be a fabulously difficult one, at least for a general Banach space. In Hilbert space the situation is more favorable to solution. Arlen Brown [16] has solved the problem of *unitary* equivalence for operators on Hilbert space satisfying $T^2 = 0$ (notice that once again we have changed the problem!).

14. VALUATION RINGS

Principal ideal rings are of precisely the right level of generality for the results we have discussed up till now. But for later purposes it will be vital to consider a more restricted class of rings. We shall not pause even to mention the general concept of valuation ring, but hasten instead to define the only kind of valuation ring that concerns us.

Definition. A discrete valuation ring R is a principal ideal ring with exactly one prime element (up to unit factors). Alternatively, we may say that R has exactly one (nonzero) prime ideal, or exactly one maximal ideal.

If R is a discrete valuation ring, the only ideals in R are of the form (p^n), where p is the unique prime element. If we take these ideals to be neighborhoods of 0 in R, we define a topology in R, making R a topological ring. If R is complete in this topology, we call it a *complete* discrete valuation ring.

We shall now proceed to explain how valuation rings occur in the theory of modules over principal ideal rings, even when the latter are not valuation rings. For this purpose, let R be an arbitrary principal ideal ring, p a prime in R. Define R_p to be the set of all elements in the quotient field of R, having the form α/β where α and β are in R and β is prime to p. It is easy to see that R_p is a discrete valuation ring with p as its unique prime. We shall write $R_p{}^*$ for the completion of R_p.

Now let M be an R-module which is primary for the prime p. There is a natural way of construing M to be a module over R_p. It comes down to the question of defining $\beta^{-1}x$ where x is in M and $\beta \in R$ is prime to p. Suppose $p^n x = 0$. We can find γ in R such that $\beta\gamma \equiv 1 \pmod{p^n}$, and we define $\beta^{-1}x = \gamma x$.

We can even get M to be an $R_p{}^*$-module; for to define $\alpha_0 x$ for α_0 in $R_p{}^*$ we choose α in R_p (or even in R) with $\alpha \equiv \alpha_0 \pmod{p^n}$ and set $\alpha_0 x = \alpha x$.

To sum up: A primary module over a principal ideal ring can just as well be regarded as a module over a discrete valuation ring, or even over a complete discrete valuation ring; it makes no difference, for the theory is identical.

It is in the theory of torsion-free and mixed modules that something new happens. Indeed, we shall see that modules over a complete discrete valuation ring are enormously simpler than modules over a general principal ideal ring; and for torsion-free modules the theory is even better than that for torsion modules.

It would be plausible to use the designation "primary module" in referring to any module over a complete discrete valuation ring, but we refrain from doing so because of possible ambiguity. Instead, we use the phrase "module over a complete discrete valuation ring" without abbreviation, and we call such a module a primary module (or, equivalently, a torsion module) only in the case where each element is annihilated by some power of p. At any rate, it should be borne in mind that there is a missing link in the theory: we have no primary decomposition such as that given by Theorem 1 for torsion modules. It might be said that the discovery of a suitable primary decomposition is the outstanding unsolved problem of the theory of torsion-free and mixed modules over a principal ideal ring.

We give two illustrations of the construction of a complete discrete valuation ring from a principal ideal ring:

(*a*) Let R be the ring of integers, and p any prime. Then R_p^* is the ring of p-adic integers; it can be regarded as the set of all series

$$a_0 + a_1 p + \cdots + a_n p^n + \cdots \qquad (0 \leqq a_i < p),$$

with addition and multiplication performed as they are in numbers written in the scale of p.

(*b*) Let R be the ring F[u] of all polynomials in u over the field F; take the prime p to be u itself. Then R_p^* is the ring of all formal power series

$$\alpha_0 + \alpha_1 u + \cdots + \alpha_n u^n + \cdots \qquad (\alpha_i \in F),$$

with the usual operations of addition and multiplication.

The two complete discrete valuation rings just constructed are typical examples. Indeed, there is a theorem of Teichmüller's showing that the second example has a very wide scope. To state it, we note that for a valuation ring R, R/(p) is a field (it is called the ***residue class field***). Teichmüller's theorem: If R is a complete discrete valuation ring, and R and its residue class field have the same characteristic, then R is isomorphic to the ring of all formal power series over its residue class field.

If R has characteristic 0 and its residue class field has characteristic different from 0 (this is the only other possibility), then R has properties reminiscent of the p-adic integers, but no complete structure theory is known.

15. TORSION-FREE MODULES

Up to this point our gaze has been directed almost entirely at torsion modules, but in this section we turn to torsion-free modules. We first prove the classical theorem on finitely generated modules. It should be noted that the torsion case of Theorem 16 has already been covered five times—in Theorems 6, 9, 11, 12, and 14. We have chosen to cite Theorem 6 at the appropriate moment in the proof of Theorem 16.

Theorem 16. Any finitely generated module over a principal ideal ring is a direct sum of cyclic modules.

Proof. This theorem has been given many proofs, some of which achieve the goal by one direct assault. But we shall be content to note how we can get it most briefly out of our previous results.

Let R be the ring, M the module. We first take up the case where M is torsion-free. Let x be one of the generators of M, and embed x in a minimal pure submodule S of M; to be explicit, S consists of all elements αx in M, where α is in the quotient field of R, such an expression having a unique meaning in a torsion-free module. By induction on the number of generators, M/S is a direct sum of cyclic modules. By Theorem 5, S is a direct summand of M. This shows that S itself is finitely generated, from which it follows readily that S is cyclic. Thus M is a direct sum of cyclic modules.

We turn now to the general case where M is no longer assumed to be torsion-free. Let T be its torsion submodule. We have just proved that M/T is a direct sum of cyclic modules. By another application of Theorem 5, T is a direct summand of M. This tells us, in turn, that T is finitely generated. By Theorem 6, T is a direct sum of cyclic modules. This concludes the proof of Theorem 16.

Next we shall generalize Theorem 13 to mixed modules, treating the torsion-free case in a special lemma. We follow usual terminology in calling an R-module *free* if it is a direct sum of copies of R.

Lemma 15. If M *is a free module over a principal ideal ring, then any submodule* N *of* M *is free.*

Remarks. (*a*) As avowed partisans of Zorn's lemma, we give a proof in that style; but we are not disputing Lefschetz's statement [D, p. 50] that in this unique case a proof by well-ordering is simpler.

(*b*) It seems noteworthy that the proof is a good deal easier here than it was in Theorem 13. This is attributable to special properties of free modules not shared by other modules (whether torsion or torsion-free). At any rate the reader will easily see that the proof of Lemma 15 cannot be simply transplanted to the torsion case.

Proof. Let $\{x_i\}$ be a basis for M. For any independent set $\{y_j\}$ in N, we form the submodule T spanned by the y's, and also the submodule S of M spanned by all the x's which appear in the expansions of the y's. We confine our attention to sets $\{y_j\}$ such that $S \cap N = T$, and we can suppose that $\{y_j\}$ is maximal among such sets. We claim that $\{y_j\}$ is a basis of N. Since $\{y_j\}$ was already assumed to be independent, we need only verify that it spans N. We suppose the contrary, and observe that any missing element of N must contain some new x's. Consider, among the missing elements of N, the subset consisting of those which involve the smallest number of new x's; then pick from this subset an element that has as coefficient of one of the new x's a ring element with the smallest available number of prime factors. Say $z = \alpha x_1 + \cdots$ is the element chosen, and α is the coefficient in question. We claim that $\{y_j\}$ can be enlarged to $\{y_j, z\}$, in contradiction of maximality. Certainly the set $\{y_j, z\}$ is still independent. Suppose w is an element of N expressible in terms of the x's appearing in $\{y_j\}$ and z. Say $w = \beta x_1 + \cdots$. The minimal choice of α shows that β must be a multiple of $\alpha, \beta = \alpha\gamma$. Then $w - \gamma z$ must lie in the submodule spanned by the y's. This shows that the set $\{y_j, z\}$ has both required properties, a contradiction.

Let us combine Theorem 13 and Lemma 15:

Theorem 17. Let R *be a principal ideal ring and* M *an* R*-module which is a direct sum of cyclic modules. Then any submodule* N *of* M *is likewise a direct sum of cyclic modules.*

Proof. Let T be the torsion submodule of M. Then $T \cap N$ is the torsion submodule of N. After a decomposition of T into primary parts, we deduce from Theorem 13 that $T \cap N$ is a direct sum of cyclic modules. Next, $N/(T \cap N)$ is in a natural way a submodule of the free module M/T. By Lemma 15, $N/(T \cap N)$ is free. Then by Theorem 5, $T \cap N$ is a direct summand of N, and thus N is a direct sum of cyclic modules.

This may be said to conclude the elementary part of the general theory of modules over principal rings. We shall now turn our attention specifically to torsion-free modules, in a systematic discussion. This discussion can (for a while at least) be carried out for an arbitrary integral domain R. We let K denote the quotient field of R.

Let M be a torsion-free R-module. We define the *rank* of M to be the maximum number of linearly independent elements—this having an unambiguous meaning. (For torsion modules or mixed modules the notion of rank is more delicate—see exercise 49.)

The rank might be regarded as the first crude invariant; to understand it better, let us take up the case where M is of rank one. Let x be any nonzero element in M. Then any y in M can be uniquely written as $y = \alpha x$ with α in K. The mapping $y \to \alpha$ is an isomorphism of M onto a set I which is an R-submodule of K (such a submodule is often called a "fractional ideal"). One easily verifies the following two statements:

(*a*) Two such modules I and J are isomorphic if and only if there exists a nonzero β in K with $J = \beta I$.

(*b*) Any module of rank one is indecomposable.

In case R is a discrete valuation ring, there are actually only two possibilities for a module of rank one: it is isomorphic either to R or K, according as it is cyclic or divisible. This is our first glimpse of the simplification that occurs on working over a valuation ring.

For an arbitrary principal ideal ring R, it is also possible to give a complete classification of the R-modules of rank one; the classification is in terms of powers of primes allowed to occur in the denominator. Full details are to be found in various papers in the literature.

This completes our discussion of modules of rank one. We turn to torsion-free modules of higher rank, and propose three questions, listing them in increasing order of optimism:

(*a*) Is every indecomposable module of rank one?
(*b*) Does every module have a direct summand of rank one?
(*c*) Is every module a direct sum of modules of rank one?

We shall see in a moment that, if R is not a complete discrete valuation ring, the first question has a negative answer; consequently all three have negative answers under the same condition. But if R is a complete discrete valuation ring, the second (and, therefore, also the first) question has an affirmative answer; and so does the third in the case of countable rank (but not otherwise).

To begin with, we shall exhibit what is probably the simplest example of an indecomposable module of rank greater than one. For the sake of concreteness we do it for ordinary groups—see exercise 50 for the generalization to modules.

Theorem 18. The additive group of p*-adic integers is an indecomposable torsion-free group whose rank is the power of the continuum.*

Proof. That the rank is the continuum follows from the fact that the cardinal number of the group is the continuum (for groups of infinite rank, the rank is the same as the cardinal number).

Now let G denote the group of p-adic integers and suppose $G = H \oplus K$. Then also $G/pG = H/pH \oplus K/pK$. But G/pG is cyclic of order p; hence, one of the summands, say H/pH, must be 0. Thus $H = pH$, and every element in H has infinite height with respect to p. This is possible only if $H = 0$, for G has no nonzero elements of infinite p-height.

We here turn our attention explicitly to modules of rank two, and prove a general existence theorem:

Theorem 19. Let R *be an incomplete discrete valuation ring. Then there exists an indecomposable torsion-free* R*-module of rank two.*

Proof. Let γ be an element in the completion of R, but not in R itself. For each n ($n = 0, 1, 2, \cdots$), pick an element δ_n in R with $\delta_n \equiv \gamma \pmod{p^n}$. Take two symbols u and v, and write $w_n = u + \delta_n v$ ($n = 0, 1, 2, \cdots$). Let M be the R-module generated by v, w_0, w_1/p, $\cdots$, w_n/p^n, $\cdots$. Let us first note that M does not contain v/p. For suppose

$$(8) \qquad \beta v + \alpha_0 w_0 + \alpha_1 w_1/p + \cdots + \alpha_n w_n/p^n = v/p$$

with δ_i and β in R. Then, in order to cancel the terms in u, we must have α_n divisible by p. Also

$$(9) \qquad w_n/p^{n-1} = w_{n-1}/p^{n-1} + (\delta_n - \delta_{n-1})v/p^{n-1}$$

and $\delta_n - \delta_{n-1}$ is divisible by p^{n-1}. On substituting equation (9) in (8), we get a similar equation, with a smaller n. Since (8) is impossible for n = 0, we ultimately reach a contradiction.

Next we claim that no element of M has infinite height. Since we have just seen that v is not even divisible by p, it will suffice for us to examine an element of the form $u + \eta v$, with η in R. But if $(u + \eta v)/p^n$ is in M, so is

$$(u + \eta v - w_n)/p^n = (\eta - \delta_n)v/p^n.$$

Since v/p is not in M, this is possible only if $\eta - \delta_n$ is divisible by p^n. This has to be true for all n. But then η is the forbidden element γ.

Suppose now that M is indeed a direct sum of modules of rank one. Each summand has to be isomorphic either to R or its quotient field. But we have just seen that M has no elements of infinite height; hence each summand must be isomorphic to R. In other words, M must be free. But now let S be the submodule of M generated by v. In M/S, the homomorphic image of u is a nonzero element having infinite height, and this is impossible if M is free. This contradiction completes the proof of Theorem 19.

Now let us turn to the other side of the question and see how far we can carry the theory of modules over a complete discrete valuation ring. In Theorem 20 we shall in fact completely settle the case of countable rank. First, a preliminary lemma:

Lemma 16. Let R *be a complete discrete valuation ring,* M *a reduced* R*-module of rank* r + 1, *and* N *a pure free submodule of* M *of rank* r. *Then* M/N *is cyclic.*

Proof. The only other possibility is that M/N is isomorphic to the quotient field of R. We shall suppose this and derive a contradiction. Let y be any nonzero element of M/N; write $y = p^n y_n$ for every n, with y_n again in M/N. The idea of the proof is to lift up y to an element x of M which will again have infinite height. First choose any z, $z_n \in M$ mapping on y, y_n, respectively, and let $u_n = p^n z_n - z$. Then u_n is in N. Also

$$u_{n+1} - u_n = p^n(pz_{n+1} - z_n).$$

By the purity of N, $u_{n+1} - u_n$ is a multiple of p^n in N. Now if we write $(\alpha_{n1}, \cdots, \alpha_{nr})$ for the components of u_n, we have that for each i, $\{\alpha_{ni}\}$

is a Cauchy sequence. By the assumed completeness, α_{ni} converges to an element β_i in R. Write $x = (\beta_1, \cdots, \beta_r)$. Then the element $z + x$ has infinite height in M, contrary to the hypothesis that M is reduced.

Theorem 20. Let R *be a complete discrete valuation ring and* M *a countably generated torsion-free* R*-module. Then* M *is the direct sum of a divisible module and a free module.*

Remark. It is an equivalent hypothesis to assume that M has countable rank.

Proof. By Theorem 3, M is the direct sum of a divisible module and a reduced module (and by Theorem 4 the divisible module is a direct sum of replicas of the quotient field of R). Thus we can assume that M is reduced.

It is evident that we can express M as the union of an ascending sequence of pure submodules $S_1 \subset S_2 \subset \cdots \subset S_i \subset \cdots$, where S_i has rank i. By induction, we assume that S_i is free. Then, by Lemma 16, S_{i+1}/S_i is cyclic. By Theorem 5, S_i is a direct summand of S_{i+1}. Then S_{i+1} in turn is free. Thus each S_j is free and is a direct summand of S_{j+1}, and this proves that M is free.

We give an example to show that the hypothesis that M is countably generated cannot be dropped in Theorem 20. The example is simply the *complete* direct sum of an infinite number of copies of R. Manifestly this can have no divisible submodules. On the other hand, the next theorem shows that it is not free (unless R is a field):

Theorem 21. Let R *be a principal ideal ring, but not a field, and let* M *be the complete direct sum of a countable number of copies of* R*. Then* M *is not free.*

Proof. To start with, we assert that M is not countably generated. For suppose, first, that R is countable. Then if M were countably generated, it would also be countable, whereas it has the power of the continuum. Suppose R is not countable. Then the elements of M given by

$$(1, \alpha, \alpha^2, \cdots),$$

with α ranging over R, all are linearly independent (as follows from consideration of the appropriate Vandermonde determinant). Hence M has uncountable rank, and cannot be countably generated. Thus if R is free, it must be free with an uncountable number of generators.

Pick a prime p in R, and let S be the submodule consisting of all $(\beta_1, \beta_2, \cdots)$ in M with the property that the power of p dividing β_n approaches infinity. If M is free, so is S, by Theorem 17. Write $x = (p, p^2, p^3, \cdots)$. Then multiplication by x is an R-isomorphism of M into a submodule of S. Hence S would also have an uncountable number of generators.

But now consider S/pS, which we can regard as a vector space over the field R/(p). It is easy to see that this vector space has countable dimension. Indeed, let e_i be the element of S with 1 in the i-th place and zeros elsewhere. Then any element of S can be written as the sum of an element in pS and a finite linear combination of the e's. In other words, the homomorphic images of the elements e_i generate S/pS (they are, in fact, a basis). Thus S/pS has countable dimension. But if S is free with an uncountable number of generators, the vector space S/pS manifestly has uncountable dimension. This contradiction completes the proof of Theorem 21.

Thus we see that Theorem 20 ceases to be valid if the countability assumption is omitted. However, there is another technique that yields at least some information in the uncountable case. It is to this that we turn in the next section.

Exercises 49-53

49. A group is said to have rank n if every finitely generated subgroup can be generated by n or fewer elements, and n is the smallest integer with this property.

(*a*) Prove that for torsion-free groups this definition agrees with the one in § 15.

(*b*) Prove that a primary group of finite rank is the direct sum of a finite number of groups, each cyclic or isomorphic to $Z(p^\infty)$.

(This concept of rank was heavily used by Prüfer in his fundamental work [91] on torsion groups.)

50. Let R be a principal ideal ring. Show that $R_p{}^*$ (the notation is that of § 14) is an indecomposable R-module. (The set-theoretic question of determining the rank of $R_p{}^*$ is not without interest.)

51. (*a*) Let M be a free module over a principal ideal ring, S a submodule of M, and T a direct summand of M. Prove that S ∩ T is a direct summand of S. (Observe that S/(S ∩ T) is isomorphic to (S + T)/T, which is free.

(*b*) Let M be a free module over a principal ideal ring. Show that the intersection of any finite number of direct summands of M is again a direct summand of M.

(*c*) Let M be a free module of countable rank over a principal ideal ring. Show that the intersection of *any* number of direct summands is a direct summand.

52. Let M be a torsion-free module of countable rank over a principal ideal ring, and suppose that every submodule of finite rank is free. Prove that M is free. (Exercises 51*c* and 52 do not hold without countability.)

53. Let M and N be torsion-free modules of rank one over a principal ideal ring; suppose that M is isomorphic to a submodule of N and that N is isomorphic to a submodule of M. Prove that M and N are isomorphic.

16. COMPLETE MODULES

Throughout this section (except in exercise 62), R will denote a discrete valuation ring, p its unique prime. Let M be an R-module with no elements of infinite height. Then in M we have a descending chain of submodules $p^i M$ with intersection 0. The fundamental idea is to use these submodules as neighborhoods of 0 for a topology in M; we shall call it the p-*adic* topology. In the usual way (Cauchy sequences mod null sequences) we form the completion M^* of M; M^* may at our pleasure be regarded again as an R-module or as a module over the completion R^* of R. If $M = M^*$ we say that M is *complete* (a designation that is to be understood as incorporating the assumption of no elements of infinite height).

Our objective in this section is to give a structure theory for complete modules, and to prove a theorem asserting that they are direct summands when they occur as pure submodules.

We first settle that M^* again has no elements of infinite height (note that the meaning of this statement does not depend on whether we are looking at M^* as an R-module or as an R^*-module).

Lemma 17. The completion, in the p-*adic topology, of a module with no elements of infinite height is again a module with no elements of infinite height.*

Proof. Let M be the module, M^* its completion, and x a nonzero element of M^* which allegedly has infinite height. We may, for instance, write $x = p^m y$ with y in M^*. Let $\{x_i\}$, $\{y_i\}$ be sequences in M converging to x and y respectively. Then $x_i - p^m y_i \to 0$; thus, for sufficiently large i, $x_i - p^m y_i$ has height $\geq$ m, and hence so does x_i. This being true for every m, we have $x_i \to 0$, $x = 0$.

It is now meaningful to assert that M^* is complete. This is proved in the obvious way: if $\{x_i\}$ is a Cauchy sequence in M^* we pick $y_i \in M$ very close to x_i. Then $\{y_i\}$ is a Cauchy sequence in M, converging say to $x \in M^*$, and x is the desired limit of $\{x_i\}$.

Lemma 17 now confronts us with the following situation: M^* has its own p-adic topology, which in turn induces a relative topology on M. It is desirable to know that the two topologies coincide. Better yet, we prove that M is a pure submodule of M^* (this implies the coincidence of the topologies, for purity says that $p^i M = M \cap p^i M^*$ for all i).

Lemma 18. A module with no elements of infinite height is pure in its p-*adic completion.*

Proof. Assume that $x \in M$ is a multiple of p^m in M^*: $x = p^m y$. We have to prove that x lies in $p^m M$. Suppose that the sequence $\{y_i\}$ in M converges to y. Then $x - p^m y_i$ approaches 0 and lies in $p^m M$ for large i.

The next lemma exhibits a useful link between the topological and algebraic aspects of R-modules.

Lemma 19. Let M *be an* R-*module with no elements of infinite height,* S *a submodule. Then* S *is dense in* M *if and only if* M/S *is divisible.*

Proof. Suppose S is dense in M. Given $x_0 \in M/S$ and an integer n, we must find $y_0 \in M/S$ with $x_0 = p^n y_0$. Lift x_0 to any $x \in M$. Since S is dense in M we can find $s \in S$ with $x - s \in p^n M$, say $x - s = p^n y$. The image y_0 of y in M/S fulfils the requirement.

Suppose M/S is divisible. Given $x \in M$ and an integer n, we must find $s \in S$ with $x - s \in p^n M$. If x_0 is the image of x in M/S we can write $x_0 = p^n y_0$ with $y_0 \in M/S$. Lift y_0 to $y \in M$; then $s = x - p^n y$ fulfils the requirement.

In proving the two main theorems we shall make systematic use of basic submodules.

Definition. Let M be any R-module. We say that S is a *basic* submodule of M if S is pure in M, S is a direct sum of cyclic modules, and M/S is divisible.

We prove a preliminary lemma and then assemble in a single lemma the facts we need about basic modules.

Lemma 20. If an R-*module is not divisible, it contains a non-zero pure cyclic submodule.*

Proof. Let M be the module, T its torsion submodule. If T is not divisible, then (Theorem 9) T contains a non-zero pure cyclic submodule, and the latter is also pure in M. So we assume T divisible, hence a direct summand of M. Since we can work within the torsion-free complement of T, we might as well assume that M itself is torsion-free. Then if x is any element of M not divisible by p, the cyclic submodule generated by x is pure.

Lemma 21. (a) Any R-*module possesses a basic submodule.*

(b) Any two basic submodules are isomorphic.

(c) Let S *be a pure submodule of the* R-*module* M, *and let* B *be a basic submodule of* S. *Then there exists a basic submodule of* M *which contains* B *as a direct summand.*

Proof. (*a*) Let $\{x_i\}$ be a maximal pure independent subset of the given module M, and let B be the submodule generated by the x's. Then B is certainly a direct sum of cyclic modules. It remains to argue that M/B is divisible. Suppose not; then (Lemma 20) M/B contains a non-zero pure cyclic submodule, generated say by y_0. Lift y_0 in any way to $y \in M$. It is routine to see that the enlarged set $\{y_i, y\}$ is still pure independent, a contradiction. (Compare ex. 22 in § 9.)

(*b*) Let B be a basic submodule of M. We must give an intrinsic description of the number of cyclic summands of a given order contained in B. We do this differently for finite and for infinite order. For finite order the fastest characterization is the following: the number of cyclic summands of order p^{n+1} in B is equal to the n-th Ulm invariant of M. This is implicit in the discussion in § 11, and we leave it to the reader to supply the details.

Let C denote the torsion-free component of B. We show the number of cyclic summands in C to be independent of the choice of B by characterizing C/pC. In fact, C/pC is isomorphic to M/(T + pM), where T is the torsion submodule of M. This will follow from the isomorphism theorem for modules when we prove that C + T + pM = M, and that $C \cap (T + pM) = pC$.

For the first of these, we note that B + pM = M since M/B is divisible. Since furthermore C + T contains B, we have C + T + pM = M.

For the second, we note that pC is obviously contained in both C and T + pM. Conversely suppose $x \in C \cap (T + pM)$, say $x = t + py$, $t \in T$, $y \in M$. Then $p^n t = 0$ for some n, hence $p^n x = p^{n+1} y$, hence $p^n x \in p^{n+1} C$ since C is pure in M. Since C is torsion-free we deduce $x \in pC$. (This argument is taken from [39, p. 200].)

(*c*) Let $\{x_i\}$ be generators of the cyclic summands of B. We have that $\{x_i\}$ is a pure independent subset of M. By adjoining $\{y_j\}$ expand it to a maximal pure independent subset of M. Then if D is the submodule spanned by the x's and y's, D is a basic submodule of M as in part (*a*). Manifestly B is a direct summand of D.

Theorem 22. Let R *be a discrete valuation ring and* M *an* R-*module with no elements of infinite height which is complete in its* p-*adic topology. Then* M *is the completion of a direct sum of cyclic modules. The number of cyclic summands of a given order is an invariant of* M; *these cardinal numbers are a complete set of invariants for* M.

Proof. Take a basic submodule B of M. We claim that M can be identified with the completion B^* of B. First we note that, since B is pure in M, the intrinsic p-adic topology in B coincides with the restriction to B of the p-adic topology in M. It follows that B^* is a submodule of $M^* = M$, and since B is dense in M (Lemma 19), $B^* = M$.

Suppose that M is the completion of another submodule B_0, where B_0 is a direct sum of cyclic submodules. By Lemma 18, B_0 is pure in M. By Lemma 19, M/B_0 is divisible. Hence B_0 is a basic submodule of M. By part (*b*) of Lemma 21, B and B_0 are isomorphic.

Theorem 23. Let R *be a discrete valuation ring,* M *an* R-*module,* S *a pure submodule. Suppose that* S *has no elements of infinite height and is complete in the* p-*adic topology. Then* S *is a direct summand of* M.

Proof. Case I. M with no elements of infinite height and complete. Let B be a basic submodule of S. By part (*a*) of Lemma 21, we enlarge B to a basic submodule $D = B \oplus C$ of M. We have that

$M = D^* = B^* \oplus C^*$ (completion of the direct sum of two modules goes component-wise). But $B^* = S$.

Case II. M with no elements of infinite height but not necessarily complete. We complete M to M^*. Since by purity the topologies match, S is a complete submodule of M^*. By Case I, S is a direct summand of M^* and, therefore, of M.

Case III. M arbitrary. Let N be the submodule of elements of infinite height. Then $S \cap N$ must be 0 (since S is pure and has no elements of infinite height). By Lemma 6, we need only show that $(S + N)/N$ is a direct summand of M/N. We note that M/N has no elements of infinite height; our proof can be completed by Case II if we show that $(S + N)/N$ is complete, and pure in M/N:

Purity. Let x_1 be in $(S + N)/N$, and suppose that $x_1 = p^r y_1$, with y_1 in M/N. On taking representatives we have

$$x = p^r y + z \qquad (x \in S,\ y \in M,\ z \in N).$$

Since $h(z) = \infty$, we have $h(x) \geq r$, $x = p^r s$ with $s \in S$. On passage modulo N, we find that x_1 is a multiple of p^r in $(S + N)/N$, as desired.

Completeness. Let $\{v_i\}$ be a Cauchy sequence in $(S + N)/N$. Choose any u_i in S mapping on v_i. For large i and j, $u_i - u_j$ has large height in M modulo N, hence actual large height in M (since elements of N have infinite height), hence large height in S (since S is pure). Thus $\{u_i\}$ is a Cauchy sequence in S and by completeness has a limit u in S. The image of u is the desired limit of the sequence v_i. This completes the proof of Theorem 23.

We remark that Theorem 23 is in a way a definitive theorem of its kind, for the assumption of completeness is inescapable. Indeed, an incomplete module is not a direct summand of its completion (exercise 56), although it is pure (Lemma 18).

If we combine Theorem 23 with Theorem 9 and the observation that a torsion-free reduced module has a pure cyclic submodule (which is of course complete), we derive:

Corollary 1. Let R *be a complete discrete valuation ring and* M *a reduced* R*-module. Then* M *has a cyclic direct summand.*

Corollary 2. Let R *be a complete discrete valuation ring with quotient field* K *and let* M *be an indecomposable* R*-module. Then* M *is isomorphic to one of the following four modules:* $R/(p^n)$, R, K, *or* K/R.

We shall conclude this section with some general remarks about the status of the theory as now developed. Consider for definiteness a reduced torsion-free R-module M (similar observations are in order for the torsion or mixed cases). If M is of countable rank, Theorem 20 shows that it is free. If M is complete, Theorem 22 shows that it is the completion of a free module. In either case we have a complete structure theory. But if we drop both the hypotheses of countability and completeness, we are able to say virtually nothing.

It seems worth while to call the reader's attention to a remarkably parallel situation in the theory of Hilbert space. There one postulates a vector space E (say over the real numbers) equipped with a bilinear inner product which is symmetric and positive-definite. This gives rise to a norm on E, and one adds the postulate of completeness. The resulting Hilbert space turns out to be uniquely determined by the specification of a single cardinal number: the number of elements in an orthonormal base. Another result, less well known, is the following: If E has countable dimension (a hypothesis incompatible with completeness), it admits an orthonormal basis in the sense of pure algebra, with finite linear combinations, only, allowed. So either countability or completeness leads to a structure theorem. If neither is assumed, the spaces seem to exist in such abundance that there is little hope of classifying them.

I suspect that a similar bleak outlook faces anyone who seeks, for example, to classify all torsion-free modules over a complete discrete valuation ring, or all primary abelian groups.

Exercises 54-62

54. Let M be an R-module with no elements of infinite height. Prove that the p-adic topology on M is discrete if and only if M is of bounded order.

55. Let M be a complete R-module, S a pure submodule. Prove that the closure of S is again pure.

56. Let M be an incomplete R-module with no elements of infinite height. Prove that M is not a direct summand of its completion.

57. Prove the finite rank case of Theorem 20 by using Theorem 23. (Hint: if R is complete so is any finitely generated R-module.)

58. Call a torsion R-module (with no elements of infinite height) torsion-complete if it is the torsion submodule of its completion. Prove the following analogue of Theorem 23: if M is a torsion R-module, S a pure torsion-complete submodule, then S is a direct summand of M.

59. Let M be a complete R-module and S a submodule of M. Prove: M/S has no elements of infinite height if and only if S is closed.

60. Let M be a complete R-module and S a closed submodule of M. Prove that M/S is complete.

61. Prove that a complete torsion R-module is of bounded order.

62. Let R be a principal ideal ring, p a prime in R. Let S be an R_p-module which is complete as such, but look at S as an R-module. Let M be an R-module containing S as a pure submodule. Prove that S is a direct summand of M. (Hint: adapt the proof of Theorem 23.)

17. ALGEBRAIC COMPACTNESS

Let R revert to a general principal ideal ring. Let M be an R-module, S a pure submodule. We have accumulated three cases where we know that S is a direct summand: (1) S divisible (Theorem 2), (2) S of bounded order (Theorem 7), (3) S a complete R_p-module (exercise 62). (Actually [2] is a special case of [3].) If we allow direct sums as well there are in fact no further examples.

Definition. A module over a principal ideal ring is *algebraically compact* (AC) if it is a direct summand of any module in which it is contained as a pure submodule.

We invite the reader to try the exercises and consult the notes at the end.

Exercises 63-65

63. Prove: a direct summand of an AC module is AC.

64. Prove: a complete direct sum of AC modules is AC.

65. Let M be AC and reduced over the principal ideal ring R. Prove that M is a complete direct sum, over the primes p in R, of modules complete over R_p.

18. CHARACTERISTIC SUBMODULES

In this section we propose to discuss the submodules of a given module M. However, we are not so ambitious as to undertake to classify all the submodules of M, though certain special cases are discussed in exercises 78-81. Rather, we shall concentrate on the characteristic and fully invariant submodules, in the sense of the following definition:

Definition. A submodule S of M is *characteristic* if it is sent into itself by every automorphism of M; S is *fully invariant* if it is sent into itself by every endomorphism of M.

We observe that a fully invariant submodule is characteristic. Below we shall see examples where the converse fails.

Though these definitions apply to arbitrary modules (or groups with operators or even more general systems), we shall deal only with primary reduced modules; as was observed in § 15, we may then just as well suppose that the ring R of operators is a complete discrete valuation ring, and we shall assume this throughout the section, except in some of the exercises.

The reader who does not yet share my enthusiasm for modules is advised that he may read this section (and the next) supposing all the modules involved to be ordinary primary abelian groups; and indeed this is the most interesting case of the theory.

Restricting our considerations to reduced modules is largely a matter of convenience; there is no serious difficulty in extending the theory to cover modules that are not reduced, as we indicate in exercises 66-73. The restriction to primary modules is another matter, and it is certainly in order at least to mention the torsion-free and mixed cases. It turns out that the torsion-free case (over a complete discrete valuation ring) is so easy to treat that we can dismiss it as exercise 72. The mixed case, on the other hand, is beyond the power of existing tools.

Our study will be based on the concept of transfinite height introduced in § 11. Let us briefly repeat the definition: We take a complete discrete valuation ring R, with prime p, and a reduced R-module M. We define the descending transfinite series $\{M_\alpha\}$ by successive multiplication by p, taking intersections at the limit ordinals. There is a first ordinal λ such that $M_\lambda = 0$, and λ is called the length of M. We set $h(x) = \alpha$ if x is in M_α but not in $M_{\alpha+1}$; and we define $h(0) = \infty$, with the understanding that ∞ exceeds any ordinal.

(A word may be inserted about the appropriate modification to make if M is not assumed to be reduced. Then the descending sequence $\{M_\alpha\}$ will end at the maximal divisible submodule M_λ. We assign to the nonzero elements of M_λ the height λ, and we stick to the convention $h(0) = \infty$. The definition of transitivity and full transitivity below makes sense, and exercises 70, 71, and 72 are to be interpreted in this way. We can also assign to M a λ-th Ulm invariant in a natural way; if P is the submodule of elements x satisfying $px = 0$, we set the λ-th Ulm invariant equal to the dimension of the vector space $P \cap M_\lambda$. The Ulm invariants, extended in just this way, are what is needed in exercise 75.)

With every element x there is associated a sequence $\{\alpha_i\}$ of ordinals given by $\alpha_i = h(p^i x)$ for $i \geqq 0$. We shall call this the *Ulm sequence* of x and write it U(x). If a second element y has $U(y) = \{\beta_i\}$, and $\alpha_i \leqq \beta_i$ for all i, we write $U(x) \leqq U(y)$. It is evident that the Ulm sequence of an element is invariant under automorphisms; also, if x is sent into y by an endomorphism, then $U(x) \leqq U(y)$. This motivates the definition of the terms "transitive" and "fully transitive" that follows:

Definition. Let R be a complete discrete valuation ring, M a reduced R-module. We say that M is *transitive* if, whenever x and y are elements of M with $U(x) = U(y)$, there exists an automorphism of M sending x into y. We say that M is *fully transitive* if, whenever x and y are elements of M with $U(x) \leqq U(y)$, there exists an endomorphism of M sending x into y.

Remark. This terminology must be accompanied by an emphatic warning: It is by no means clear that full transitivity implies transitivity. As far as is known, the two concepts may be independent.

For torsion-free modules it is easy to prove that any R-module is both transitive and fully transitive (see exercise 72). For primary modules we can prove both with the aid of a hypothesis of "local countability." For the proof of Theorem 24, we simply cite exercises 38 and 39 of §11.

Theorem 24. Let R *be a complete discrete valuation ring and* M *a primary* R*-module with the property that any two elements of* M *can be embedded in a countably generated direct summand of* M. *(This is in particular the case if* M *itself is countably generated, or if it has no elements of infinite height.) Then* M *is both transitive and fully transitive.*

This is as far as we are able to go in proving that modules are transitive. We turn now to investigating the consequences of transitivity. In the remainder of the section M will always be a reduced primary module over the complete discrete valuation ring R (this will be restated in the theorems but not in the definitions or lemmas).

Consider the Ulm sequence $U(x) = \{\alpha_i\}$ of an element x; note that if x is of finite order, the sequence is ∞ from some point on. For any n such that $\alpha_{n-1} \neq \infty$ we have $\alpha_{n-1} < \alpha_n$. It may happen that $\alpha_{n-1} + 1 < \alpha_n$; for instance, this will certainly be the case if $\alpha_n = \infty$. We shall then say that a *gap* occurs between α_{n-1} and α_n. Lemma 22 indicates that there is a relation between the presence of gaps and the Ulm invariants of M; a similar phenomenon was encountered in the proof of Theorem 14.

Lemma 22. Let $\{\alpha_i\}$ *be the Ulm sequence of an element* x *in* M, *and suppose that there is a gap between* α_{n-1} *and* α_n. *Then the* α_{n-1}*-th Ulm invariant of* M *is nonzero.*

Proof. Since $h(p^n x) = \alpha_n > \alpha_{n-1} + 1$, we can write $p^n x = py$ with $h(y) > \alpha_{n-1}$. Set $z = p^{n-1}x - y$; then $pz = 0$ and $h(z) = \alpha_{n-1}$. This shows that the α_{n-1}-th Ulm invariant is nonzero.

Motivated by Lemma 22, we introduce a definition:

Definition. Let $\{\alpha_i\}$ be a monotone increasing sequence of ordinals defined for $i \geqq 0$; suppose that each α_i is less than λ (the length of M) except that it is permitted that the sequence be ∞ from some point on. We call $\{\alpha_i\}$ a *U-sequence* (relative to M) if, whenever a gap occurs between α_{n-1} and α_n, the α_{n-1}-th Ulm invariant of M is nonzero.

Lemma 22 may now be restated as follows: Any Ulm sequence is a U-sequence. In Lemma 24 we shall see that the converse holds for those U-sequences that end in ∞'s. We first prove the following preparatory lemma:

Lemma 23. Let x *be an element of height* α, *and let* β *be an ordinal less than* α. *Then (a) if* $\beta + 1 = \alpha$, *we can write* $x = py$ *with* $h(y) = \beta$; *(b) if* $\beta + 1 < \alpha$, *we can write* $x = pz$ *with* $h(z) > \beta$; *(c) if the* β*-th Ulm invariant of* M *is nonzero, we can write* $x = py$ *with* $h(y) = \beta$.

Proof. (*a*) If $\beta + 1 = \alpha$, we have $M_\alpha = pM_\beta$. Then x can be written as py with y in M_β, and y necessarily has height β.

(*b*) Since $\alpha \geqq \beta + 2$, x lies in $M_{\beta+2}$ and can be written pz with z in $M_{\beta+1}$.

(*c*) If $\beta + 1 = \alpha$, we cite part (*a*). Otherwise we write $x = pz$, as in part (*b*). Since the β-th Ulm invariant is nonzero, there exists an element w of order p and height β. We set $y = w + z$.

Lemma 24. Let $\{\alpha_i\}$ *be a* U*-sequence with* $\alpha_n \neq \infty$ *and* $\alpha_{n+1} = \infty$. *Then there exists an element* x *in* M *with* $U(x) = \{\alpha_i\}$. *Moreover,* x *can be chosen so that* $p^n x$ *is an arbitrarily prescribed element of order* p *and height* α_n.

Proof. Since there is a gap between α_n and α_{n+1}, the α_n-th Ulm invariant of M is nonzero. Hence there exist elements of order p and height α_n. Let z be a prescribed element of this form. By n successive applications of Lemma 23, using part (*a*) or part (*c*) as required, we write $z = p^n x$ with $U(x) = \{\alpha_i\}$.

A U-sequence which does not end in ∞'s cannot be the Ulm sequence of any element in a primary module. But it is possible to match a given U-sequence at one coördinate, while equaling or exceeding it everywhere.

Lemma 25. Let $\{\alpha_i\}$ *be a* U*-sequence and* n *a given nonnegative integer. Then there exists in* M *an element* x *with* $U(x) \geqq \{\alpha_i\}$, $h(p^n x) = \alpha_n$.

Proof. Suppose first that a gap between α_m and α_{m+1} occurs for some $m \geqq n$. Then the sequence

$$\alpha_0, \alpha_1, \cdots, \alpha_n, \cdots, \alpha_m, \infty, \cdots$$

is a U-sequence, and by Lemma 24 there exists an element x having it as its Ulm sequence. It remains to treat the case where there are no gaps from α_n on. We then take any element z of height α_n and, as in the proof of Lemma 24, descend to an element x by n applications of

Lemma 23. If $U(x) = \{\beta_i\}$, then $\beta_i = \alpha_i$ for $i \leqq n$; for $i > n$ we must have $\beta_i \geqq \alpha_i$, since the α's increase by only one at every step.

The U-sequences relative to M form a partially ordered set under the natural pointwise ordering, and in this partially ordered set greatest lower bounds exist.

Lemma 26. The pointwise inf of any number of U-*sequences is also a* U-*sequence.*

Proof. Let $\{\beta_i\}$ be the inf in question and suppose that there is a gap between β_{n-1} and β_n. Among the sequences of which we are taking the inf there must be one, say $\{\alpha_i\}$, such that $\alpha_{n-1} = \beta_{n-1}$. Necessarily $\alpha_n \geqq \beta_n$, so that there is also a gap between α_{n-1} and α_n, and hence the α_{n-1}-th Ulm invariant of M is nonzero. This shows that $\{\beta_i\}$ is a U-sequence.

In the partially ordered set of U-sequences there is a largest element, namely, the sequence consisting entirely of ∞'s. It follows that least upper bounds also exist, and the U-sequences form a complete lattice. But one must beware—the sup in this lattice is not necessarily taken pointwise (see exercise 74).

For any U-sequence $\{\alpha_i\}$ we define $M\{\alpha_i\}$ as the set of all x in M with $U(x) \geqq \{\alpha_i\}$. It is evident that $M\{\alpha_i\}$ is a fully invariant submodule of M. Conversely, let a fully invariant submodule S be given. We define α_i to be the inf of $h(p^i x)$, taken over all x in S. By Lemma 26 the resulting sequence $\{\alpha_i\}$ is a U-sequence; we call it the Ulm sequence of S and write it U(S). It is an immediate consequence of Lemma 25 that $\{\alpha_i\} = U(M\{\alpha_i\})$ for any U-sequence $\{\alpha_i\}$. If M is fully transitive, it is also true that $S = M\{U(S)\}$ for any fully invariant submodule S. The inclusion of S in $M\{U(S)\}$ is clear. To prove the reverse inclusion we let w be an element of $M\{U(S)\}$; we have to show that w is in S. Write $U(w) = \{\gamma_i\}$ and suppose that the order of w is p^{n+1}. There must exist in S an element y whose Ulm sequence $\{\beta_i\}$ satisfies $\beta_n \leqq \gamma_n$. We can construct ordinals $\delta_0, \cdots, \delta_{n-1}$ satisfying $\delta_i \geqq \beta_i$ and such that

$$\delta_0, \delta_1, \cdots, \delta_{n-1}, \gamma_n, \infty, \cdots$$

is a U-sequence. By Lemma 24 there exists an element x having this as its Ulm sequence and satisfying $p^n x = p^n w$. Since S is fully invariant, M is fully transitive, and $U(x) \geqq U(y)$, we have $x \in S$. Since, further, $w - x$ is in $M\{U(S)\}$ and has order less than p^{n+1}, we may assume by induction that it lies in S. Hence $w \in S$.

We have thus established a one-to-one correspondence between fully invariant submodules and U-sequences. The correspondence is order-inverting and hence is a lattice anti-isomorphism. Since finite sups and infs of sequences are taken pointwise, it follows that both lattices are distributive. In fact, since arbitrary infs of sequences are taken pointwise, we have the appropriate infinite distributive law; dualized to submodules it reads:

$$A \cap (\cup B_i) = \cup(A \cap B_i). \tag{10}$$

However, for the other infinite distributive law,

$$A \cup (\cap B_i) = \cap(A \cup B_i), \tag{11}$$

the argument fails, and indeed the result itself is false—see exercise 74.

Let us summarize our information in a theorem:

Theorem 25. Let R *be a complete discrete valuation ring and* M *a fully transitive reduced primary* R*-module. Then any fully invariant submodule of* M *has the form* $M\{\alpha_i\}$, *where* $\{\alpha_i\}$ *is a* U*-sequence relative to* M*. This correspondence is one-to-one, and implements an anti-isomorphism between the lattice of fully invariant submodules and the lattice of* U*-sequences. Both lattices are distributive. The lattice of fully invariant submodules even satisfies the infinite distributive law, equation* (10), *but not necessarily its dual, equation* (11).

Our final topic will be the circumstances under which we can deduce that transitivity implies full transitivity, or that characteristic submodules are fully invariant. The topic derives additional interest from the existence of a bizarre exceptional case. The two definitions and five lemmas that follow are preludes to the proof of Theorem 26.

Definition. Let α be an ordinal. We say that M has *property* $P(\alpha)$ if, for any element x of order p and height α, we can find an element y such that both y and x + y have order p and height α.

Lemma 27. If M *fails to have property* $P(\alpha)$, *then the residue class field of* R *has two elements and the* α*-th Ulm invariant of* M *is one.*

Proof. Suppose that on the contrary the residue class field of R has more than two elements. Then there exists in R an element c which is not congruent to 0 or -1 mod p. Given x of order p and height α, we take y = cx. Since neither c nor 1 + c is divisible by p, both y and x + y have height α. This contradicts our hypothesis that M fails to have property $P(\alpha)$.

The α-th Ulm invariant counts the number of elements of order p and height α which are linearly independent modulo those of greater height. If we are given one such element x, and the α-th Ulm invariant is not one, we can take for y any linearly independent element of order p and height α.

Definition. Let x and t be elements of M, with Ulm sequences $\{\alpha_i\}$ and $\{\gamma_i\}$. Suppose that the order of t is p^{k+1}. We say that t is *normal relative* to x if it satisfies the following two conditions: (*a*) $U(t) \geq U(x)$, (*b*) there exists an integer $r \leqq k$ such that the sequence $\gamma_r, \cdots, \gamma_k$ is free from gaps and $\gamma_i > \alpha_i$ for $i \leqq r - 1$.

Lemma 28. Suppose that t *has order* p^{k+1} *and is normal relative to* x. *Let* $\{\alpha_i\}$, $\{\gamma_i\}$, *and* $\{\delta_i\}$ *be the Ulm sequences of* x, t, *and* x + t. (*a*) *If* $\delta_k = \alpha_k$, *then* $U(x + t) = U(x)$. (*b*) *If* $\delta_k > \alpha_k$, *there is a gap between* α_k *and* α_{k+1}.

Proof. The sequence $\{\delta_i\}$ necessarily coincides with $\{\alpha_i\}$ for $i \leqq r - 1$ and for $i > k$. Suppose that $\delta_j > \alpha_j$ for some j with $r \leqq j \leqq k$. Then $\gamma_j = \alpha_j$, $\delta_j > \gamma_j$, and, since the γ's rise by just one at each step, $\delta_k > \gamma_k \geqq \alpha_k$. If $\delta_k = \alpha_k$, as assumed in part (*a*). we must therefore have $\delta_i = \alpha_i$ for all i, and $U(x + t) = U(x)$. If $\delta_k > \alpha_k$, as in part (*b*), then $\alpha_{k+1} = \delta_{k+1} > \delta_k > \alpha_k$, and consequently there is a gap between α_k and α_{k+1}.

Lemma 29. Let x *be an element of order* p^{n+1}, *and let* $\{\alpha_i\}$ *be its Ulm sequence. Let* r *be the smallest integer such that the sequence* $\alpha_r, \cdots, \alpha_n$ *is free from gaps. Then we can write* x = v + w *in such a way that*

$$\alpha_0, \alpha_1, \cdots, \alpha_{r-1}, \infty, \cdots$$

is the Ulm sequence for v, w *is normal relative to* x, w *has order* p^{n+1}, *and* $h(p^n w) = \alpha_n$.

Proof. By r successive applications of part (*b*) of Lemma 23, we write $p^r x = p^r w$ with $h(p^i w) > \alpha_i$ for $i \leqq r - 1$. On setting v = x - w, we fulfill the requirements of the lemma.

Let us change notation in Lemma 29, replacing n, x, v, and w by k, y, s, and t. If we are further given an element x with $U(x) \leqq U(y)$, t is normal relative to x as well as to y. The result is the following lemma:

Lemma 30. Let x *and* y *be elements with* $U(x) \leqq U(y)$, *and let* p^{k+1} *be the order of* y. *Then we can write* y = s + t *in such a way that* $U(s) \geqq U(x)$, s *has smaller order than* p^{k+1}, t *has order* p^{k+1}, $h(p^k t) = h(p^k y)$, *and* t *is normal relative to* x.

Lemma 31. Let x *be an element with Ulm sequence* $\{\alpha_i\}$. *Suppose that* t *is normal relative to* x, *has order* p^{k+1}, *and satisfies* $h(p^k t) = \alpha_k$. *Suppose further that* M *is transitive and has the property* $P(\alpha_k)$. *Then* t *lies in the characteristic submodule* N *generated by* x.

Proof. If $h(p^k x + p^k t) = \alpha_k$, then $U(x + t) = U(x)$ by part (*a*) of Lemma 28. Since M is transitive, this implies that x + t is in N, and hence that t is in N. We may therefore suppose that $h(p^k x + p^k t) > \alpha_k$. Since M has the property $P(\alpha_k)$, we can supplement $p^k t$ with a second element g of order p and height α_k in such a way that $h(p^k t - g) = \alpha_k$. By Lemma 24 we write $g = p^k t'$ with $U(t') = U(t)$. The height of $p^k x + p^k t'$ must be α_k, for otherwise $h(p^k t - g)$ would exceed α_k. The argument above, applied now to t' instead of t, yields $t' \in N$. By the transitivity of M, $t \in N$.

Theorem 26. Let R *be a complete discrete valuation ring and* M *a transitive reduced primary* R*-module. Assume that either of the following holds:*

(*a*) *The residue class field of* R *has more than two elements;*

(*b*) M *has at most two Ulm invariants equal to one, and if it has exactly two, they correspond to successive ordinals.*

Then: M *is fully transitive, and its characteristic submodules are fully invariant.*

Proof. Let x and y be elements of M with Ulm sequences $\{\alpha_i\}$ and $\{\beta_i\}$, and suppose that $U(x) \leqq U(y)$. Let N denote the characteristic submodule of M generated by x. We shall prove that y is in N; this will imply both that M is fully transitive and that characteristic submodules of M are fully invariant. The proof will be by induction on n + k, where p^{n+1} and p^{k+1} are the orders of x and y.

We begin by applying Lemma 30, using the same notation as in that lemma. The element s is in N by induction, and so it remains to prove that t is in N. Write $z = x + t$. Since $U(t) \geqq U(x)$, we have $U(z) \geqq U(x)$, and in particular $h(p^k z) \geqq \alpha_k$. If $h(p^k z) = \alpha_k$, then $U(z) = U(x)$ by part (*a*) of Lemma 28. From this it follows (by the transitivity of M) that z is in N. We may therefore assume that $h(p^k z) > \alpha_k$. We note that by part (*b*) of Lemma 28 there is a gap between α_k and α_{k+1}; this fact will be held in reserve for future use. Since $p^k z = p^k x + p^k t$, a sum in which the heights are α_k and β_k, we must have $\alpha_k = \beta_k$. Let us note particularly the following fact: For $\beta_k > \alpha_k$ we have proved that y is in N.

At this point a special argument is possible for the case k = n. If the order of z is smaller than p^n, then by induction z lies in N. Otherwise we let z assume the role hitherto occupied by y. Since $h(p^n z) > \alpha_n$, the result obtained at the end of the preceding paragraph is applicable to z, and yields $z \in N$. Henceforth we therefore assume $k < n$. This implies $\alpha_n > \alpha_k + 1$, for, as we noted above, there is a gap between α_k and α_{k+1}. It follows from Lemma 27 and our hypothesis that M must have either the property $P(\alpha_k)$ or the property $P(\alpha_n)$. If M has the property $P(\alpha_k)$, t lies in N by Lemma 31. If M has the property $P(\alpha_n)$, we perform on x the decomposition given by Lemma 29. We note that w is in N by Lemma 31, and hence so is v. The integer r is at most k, for there is a gap between α_k and α_{k+1}, whereas the sequence $\alpha_r, \cdots, \alpha_n$ is free from gaps. Hence $U(v) \leqq U(t)$. Since the order of v is less than p^{n+1}, we can apply induction again to deduce that t lies in the characteristic submodule generated by v. But $v \in N$, and therefore $t \in N$. This completes the proof of Theorem 26.

Theorem 27. Let R *be a complete discrete valuation ring with a residue class field having two elements. Let* M *be a fully transitive reduced primary* R*-module, and suppose there exist ordinals* α, β, *with*

$\beta > \alpha + 1$, *such that the* α*-th and* β*-th Ulm invariants are both one. Then* M *has a characteristic submodule that is not fully invariant.*

Proof. The desired submodule S is generated by all elements x satisfying $p^2x = 0$, $h(x) = \alpha$, $h(px) = \beta$, or, in other words, elements whose Ulm sequence is $(\alpha, \beta, \infty, \cdots)$. That S is characteristic is clear. If S is to be fully invariant it must (by the full transitivity of M) contain an element of order p and height α. But the general element y of S has the form $y = \Sigma c_i x_i$, where $c_i \in R$ and $U(x_i) = (\alpha, \beta, \infty, \cdots)$. Also, each $c_i \equiv 0$ or 1 (mod p). There are two possibilities: If an odd number of c's are $\equiv 1$, then $U(y) = (\alpha, \beta, \infty, \cdots)$. If an even number are $\equiv 1$, then $h(y) > \alpha$.

Exercises 66-81

66. Let R be any integral domain and M a torsion-free divisible R-module. Show that:

(*a*) If x and y are any nonzero elements of M, there exists an automorphism of M sending x into y.

(*b*) The only characteristic submodules of M are 0 and M.

67. Let R be a principal ideal ring and M a torsion R-module. Show that any characteristic (fully invariant) submodule of M is a direct sum of characteristic (fully invariant) submodules of the primary components of M.

68. Let R be a principal ideal ring and M a divisible primary R-module. Show that the only characteristic submodules of M are the submodules consisting of all elements x satisfying $p^i x = 0$ $(i = 0, 1, \cdots)$, and that they are fully invariant.

69. Let R be a principal ideal ring, M a divisible R-module, and T its torsion submodule. Show that a characteristic submodule of M is either all of M or else is a characteristic submodule of T, and that it is fully invariant.

70. Let R be a discrete valuation ring and M a divisible R-module. Show that M is both transitive and fully transitive.

71. Let R be a discrete valuation ring, M an R-module, and D its maximal divisible submodule.

(*a*) Prove that M is transitive if and only if M/D is transitive.

(*b*) Prove that M is fully transitive if and only if M/D is fully transitive. (The one thing that is not routine is the following: Let x and y be such that x - y is in D, and the cyclic submodules S and T which they generate are disjoint from D. To get an automorphism sending x into y, expand S and T to be maximal with respect to disjointness from D; observe that you get direct summands.)

72. Let R be a complete discrete valuation ring, M a torsion-free R-module, D its maximal divisible submodule.

(*a*) Prove that M is both transitive and fully transitive.

(*b*) Prove that the only characteristic submodules of M are 0, D, and $p^i M$ $(i = 0, 1, \cdots)$, and that they are fully invariant.

(The key is Theorem 23, which assures us that any pure submodule of M of finite rank is a direct summand.)

73. Let R be a complete discrete valuation ring, M a primary R-module, and D its maximal divisible submodule, so that $M = D \oplus E$. Then show that any fully invariant submodule S of M has the form $S = T \oplus U$, where T and U are fully invariant submodules of D and E. Show also that T = D unless U is of bounded order and that if U is annihilated by precisely p^i, so is T. (Observe that a corresponding statement for characteristic submodules fails, in the light of exercise 75.)

74. Let M be a primary module with nonvanishing Ulm invariants, at least as far as the ordinal $\omega + 1$. Let B_i and A be the fully invariant submodules determined by the U-sequences $(i, i + 1, \infty, \cdots)$ and $(1, \omega + 1, \infty, \cdots)$.

(*a*) Observe that the sup of the sequences for B_i is $(\omega, \omega + 1, \infty, \cdots)$ and so is not taken pointwise.

(*b*) Show that the infinite distributive law, equation (11), fails.

75. Show that Theorems 26 and 27 remain valid even if M is not reduced (provided that the Ulm invariants are appropriately interpreted, as observed above).

76. Let R be a discrete valuation ring and M an R-module with no elements of infinite height. Show that the sup of U-sequences is taken pointwise, and that consequently the lattice of U-sequences satisfies both distributive laws.

77. Consider a fully invariant submodule S of the form $M\{\alpha_i\}$.

(*a*) Show that there always exists a countable set spanning S in the sense that S is the smallest fully invariant submodule containing them.

(*b*) Show that if S is finitely spanned in this sense there always exists a single element spanning S, and, in particular, that this is always the case if M is of bounded order.

78. (*a*) Let M be a module over a discrete valuation ring. Call a submodule S *regular* if

$$p^n S \cap p^{n+k} M = p^n (S \cap p^k M)$$

holds for all n, k. Note that inequality always holds in one direction. Prove that a pure submodule is regular. (This concept of regularity was introduced by Vilenkin [122].)

(*b*) Suppose further that M is of bounded order. Show then that a submodule S is regular if and only if it is possible to choose "simultaneous bases" for M and S, i.e., a basis $\{x_i\}$ of M such that suitable multiples $\{c_i x_i\}$ constitute a basis for S. (Note that there is no corresponding result if M is not of bounded order, even if S is pure—see exercise 16, in § 8.)

79. (*a*) Show that the following statements are equivalent for a primary module M:

(1) $h(px) = ph(x)$ for every nonzero x in M.

(2) M has at most two nonvanishing Ulm invariants and they correspond to consecutive integers.

(3) For a suitable integer n, M is a direct sum of cyclic modules of orders p^n and p^{n+1}.

(4) Every submodule of M is regular.

(*b*) Observe that M satisfies any (and hence all) of these conditions if $p^2M = 0$.

80. (*a*) Let R be a principal ideal ring, M a free R-module, S a submodule containing αM for some nonzero α in R. Show that it is possible to choose simultaneous bases for M and S.

(*b*) Observe that simultaneous bases can exist only if M/S is a direct sum of cyclic modules. (I have not been able to determine whether this condition is also sufficient.)

81. Let M be a primary module which is a direct sum of cyclic modules of a fixed order p^n. Show that the following two statements are equivalent for submodules S, T of M:

(*a*) There exists an automorphism of M carrying S into T.

(*b*) S and T are isomorphic, and M/S and M/T are isomorphic.

19. THE RING OF ENDOMORPHISMS

We start with a remark like that at the beginning of § 18. We shall be studying primary modules, which may as well be taken over a complete discrete valuation ring. The reader may stick to the case of ordinary primary groups, but he is advised that this time there is an interesting application to linear transformations.

The project is to discover the extent to which a module can be recaptured from its ring of endomorphisms. In order to reach the heart of the matter as rapidly as possible, we shall leave to the reader as exercises several subsidiary questions.

Let R be a complete discrete valuation ring, M a primary R-module. We shall write A = E(M) for its ring of endomorphisms. Some of these endomorphisms are induced by the elements of R itself: to $\alpha \in R$ we make correspond the endomorphism that sends every x into αx. In this way we get at any rate a homomorphism of R into E(M). We shall explicitly concern ourselves only with the case where this mapping is an isomorphism, or where, in other words, no nonzero element of R annihilates all of M; one then speaks of M as being a *faithful* R-module. We can simply regard R as being embedded in E(M); it will be comforting later (Theorem 29) to learn that it is even the center of E(M).

For the contrary case where M is not a faithful R-module, see exercise 88.

Let N be a second faithful primary R-module, and B = E(N) its ring of endomorphisms. Then both A and B contain R, and we can speak of an isomorphism from A to B being the identity on R; we then call it an R-isomorphism. The complication that occurs in studying other isomorphisms is not of great interest—see exercise 89.

Let T be an isomorphism between M and N. Then the mapping $S \to T^{-1}ST$ is an isomorphism between E(M) and E(N); we call it the isomorphism *induced* by T.

We are now ready to state and prove the main theorem.

Theorem 28. Let R *be a complete discrete valuation ring,* M *and* N *faithful primary* R*-modules. Then any* R*-isomorphism between* E(M) *and* E(N) *is induced by an isomorphism of* M *and* N.

Proof. Let D be the maximal divisible submodule of M, and write $M = D \oplus F$. It proves vital to distinguish two cases:

Case I. F *is not of bounded order.* By iterated use of exercise 27, of § 9, we can find submodules S_i, T_i ($i = 1, 2, \cdots$) with the following properties:

(*a*) Each S_i is cyclic of order $p^{n(i)}$, with n(i) increasing monotonically,

(*b*) $T_1 \supset T_2 \supset T_3 \supset \cdots$,

(*c*) $S_j \subset T_i$ for $j > i$,

(*d*) M is the direct sum

$$S_1 \oplus S_2 \oplus \cdots \oplus S_i \oplus T_i.$$

Let x_i be a generator of S_i. Let e_i be the element of A = E(M) which is the projection of M on the direct summand S_i (the other summands T_i, $S_1, S_2, \cdots, S_{i-1}$ being annihilated by e_i). For $i < j$, define e_{ji} to be the endomorphism that sends x_j into x_i and annihilates the complement of S_j; also define e_{ij} to be the endomorphism sending x_i into $p^{n(j)-n(i)}x_j$, and annihilating the complement of S_i. These elements are very much like a set of matrix units. In detail we have:

$$e_i^2 = e_i,\ e_ie_j = e_je_i = 0 \qquad (i \neq j) \tag{12}$$

$$e_ie_{ij} = e_{ij}e_j = e_{ij} \tag{13}$$

$$e_{ij}e_{jk} > e_{ik} \text{ for } i < j < k \text{ or } i > j > k \tag{14}$$

$$e_{ij}e_{ji} = p^{|n(j)-n(i)|}e_i. \tag{15}$$

Now let U be the given isomorphism between A = E(M) and B = E(N). Write $e_iU = f_i$, $e_{ij}U = f_{ij}$. Then U maps e_iAe_i isomorphically on f_iBf_i. It follows from exercises 83 and 84 that Nf_i is a cyclic direct summand of N and has order $p^{n(i)}$. We propose to select inductively generators y_i of Nf_i so as to satisfy

$$y_{i+1}f_{i+1,i} = y_i. \tag{16}$$

Suppose this has been done as far as y_i. First take any generator z of Nf_{i+1}. In view of the equation

$$f_{i+1,i}f_i = f_{i+1,i},$$

which follows from (13), we have that $zf_{i+1,i}$ lies in Nf_i, and so is of the form αy_i ($\alpha \in R$). We claim that α is a unit, for otherwise we would get a contradiction on applying the equation derived from (15),

$$f_{i+1,i}f_{i,i+1} = p^{n(i+1)-n(i)}f_{i+1},$$

to the element z (the right side would have order $p^{n(i)}$ and the left side smaller order). It remains only to define $y_{i+1} = \alpha^{-1}z$, and we satisfy (16). It follows further that for all $i < j$ we have

(17) $$y_j f_{ji} = y_i.$$

We are now ready to define the isomorphism (we shall call it V) between M and N which is to induce the isomorphism U of E(M) and E(N). Given an element x in M, we write it as $x = x_i C$ for C in E(M); note that this can be done for any i large enough so that $p^{n(i)}$ exceeds the order of x. Then, writing C' for CU, we define $xV = y_i C'$. Four things need to be verified:

(*a*) It is most urgent to check that the definition is independent of the choice of i and C. So suppose x is also $x_j G$ for G in E(M). We can assume $j \geqq i$. According as $j = i$ or $j > i$ we have that $C - G$ or $e_{ji}C - G$ annihilates x_j and hence e_j. If we take $e_{ji} = e_j$ in case $j = i$, we can write $e_{ji}C = e_j G$ uniformly. Apply the isomorphism U that carries e(M) to E(N) and we get $f_{ji}C' = f_j G'$; and on applying each of these to y_j, we get, by equation (17), $y_i C' = y_j G'$. This tells us that the two proposed images of x coincide, and thus V is well defined.

(*b*) To see that V is a homomorphism we first check additivity. Given x and z in M, we can for suitably large i write both x and z in terms of x_i:

$$x = x_i C, \; z = x_i G, \; C, G \in E(M).$$

Then (in the same notation as above) we have by definition:

$$xV = y_i C', \; zV = y_i G', \; (x + z)V = y_i (C' + G').$$

Hence $(x + z)V = xV + zV$. The proof that V preserves scalars (elements of R) is similar.

(*c*) It is immediate that V is one-to-one and onto.

(*d*) It remains finally to check that V induces the isomorphism U. In other words, given S in E(M) we have to verify

(18) $$SU = V^{-1}SV.$$

Now (18) need be verified only when applied to a typical element of N;

and we can take this typical element to be $y_iC' = (x_iC)V$. Then the right side of (18) becomes x_iCSV; by definition of V, this in turn is $y_iC'S' = y_i(CS)U$, as desired.

Case II. F *is of bounded order.* The proof is quite similar to that of Case I; the main difference is that here we must take special account of the elements of D. Since the argument is so similar we shall merely outline it.

Pick an element x of maximal order in F; it will generate a cyclic direct summand of F, and we let e denote the projection on this summand. Again, pick a direct summand of D isomorphic to $Z(p^\infty)$; let e_1 be the projection on it; and let x_i ($i = 1, 2, \cdots$) be the usual generators of $Z(p^\infty)$, with

$$px_1 = 0, \; px_2 = x_1, \cdots .$$

The given isomorphism U sends e and e_1 into, say, f and f_1, where f projects on a cyclic direct summand of the same order as that generated by x, and f_1 projects on a direct summand isomorphic to $Z(p^\infty)$. We pick a generator y for the first, and generators y_i for the second with

$$py_1 = 0, \; py_2 = y_1, \cdots .$$

Now consider an arbitrary element t of M. First it has a unique expression $t = t_1 + t_2$ in the direct sum $M = D \oplus F$. Next we can find C in E(M) sending x into t_2 and x_i (for i sufficiently large) into t_1, and annihilating the complementary summands. We define $tV = (y_i + y)C'$, where C' is the image of C under U. We shall conclude by showing that this definition is a sound one (the rest of the argument is virtually unchanged from Case I). Suppose, then, that also $t = (x_j + x)G$, where $j \geqq i$. We find that $C - G$ annihilates x, while $p^{j-i}C - G$ annihilates x_j. This gives us that $e(C - G) = 0$, and also gives (by exercise 85) that $e_1(p^{j-i}C - G)$ is a multiple of p^j in the ring e_1Ae_1. Now apply U, and we find that $f(C' - G') = 0$ and that $f_1(p^{j-i}C' - G')$ is a multiple of p^j in f_1Bf_1. From these facts in turn we deduce $y(C' - G') = 0$, $y_j(p^{j-i}C' - G') = 0$, and so

$$(y_i + y)C' = (y_j + y)G'.$$

These are precisely the two proposed images of t. We have concluded the proof of Theorem 28.

Next we identify the center of a ring of endomorphisms:

Theorem 29. Let R *be a complete discrete valuation ring and* M *a faithful primary* R*-module. Then the center of the ring of endomorphisms of* M *is precisely* R.

Proof. We follow the notation used in the proof of Theorem 28, as well as the division into two cases.

Case I. Let C be in the center of E(M). From the fact that C commutes with e_i we get at once that x_iC is a multiple of x_i, say $x_iC = \alpha_i x_i$. Now take the equation

$$e_{ji}C = Ce_{ji} \qquad (j > i)$$

and apply it to x_j. The result is

$$\alpha_i x_i = \alpha_j x_j e_{ji} = \alpha_j x_i,$$

whence $\alpha_j \equiv \alpha_i \pmod{p^{n(i)}}$. In this way a unique element α of R is determined, congruent to each α_i mod $p^{n(i)}$. We claim that $xC = \alpha x$ holds for an arbitrary x in M. To see this, write $x = x_iG$ and observe

$$xC = x_iGC = x_iCG = \alpha_i x_iG = \alpha x_iG = \alpha x.$$

Case II. We begin by using the fact that C commutes with e_1. This means that C leaves invariant the submodule Me_1, which is isomorphic to $Z(p^\infty)$. So C induces an endomorphism on Me_1, which, according to exercise 82, is given by a certain element α in R.

Next let us prove that $xC = \alpha x$. We observe that C commutes with e, which at least tells us that $xC = \beta x$ for some β in R. Let x_i be the element in Me_1 with the same order as x. We can construct an endomorphism H sending x into x_i. Then

$$xCH = \beta xH = \beta x_i = xHC = x_iC = \alpha x_i.$$

Hence β agrees with α, modulo the order of x_i. But this means $\beta x = \alpha x$.

We are ready to take on the general element t of M, and to show that $tC = \alpha t$. We can write $t = (x_j + x)G$ for G in E(M). From CG = GC we find

$$(x_j + x)CG = \alpha(x_j + x)G = \alpha t$$

$$= (x_j + x)GC = tC.$$

The next theorem is nothing but a corollary of Theorem 28. We delayed it till after Theorem 29 in order to be able to refer conveniently to the center of E(M). Also, we can drop the requirement that M be faithful—see exercise 88.

Theorem 30. Let R *be a complete discrete valuation ring,* M *a primary* R*-module. Then any automorphism of* E(M) *which leaves the center elementwise fixed is inner.*

Our final topic is an application of Theorem 29 to linear transformations. We review briefly the setup which was studied in § 12. V is a vector space over a field F, T a linear transformation on V. We construe V to be a module over F[u] with u an indeterminate. We shall suppose V to be faithful as an F[u]-module; the contrary case is that where T is algebraic—see exercise 90.

Also, to simplify matters we assume T to be locally nilpotent. Then V is a primary module over F[u]. In accordance with our standard practice we extend F[u] to be a complete discrete valuation ring, namely, the ring R of all formal power series in u with coefficients in F.

The reader will have no difficulty in verifying the following two statements.

(*a*) The ring of endomorphisms of the module V is the set of all linear transformations commuting with T.

(*b*) The center of the ring of endomorphisms is the set of all linear transformations that commute with every linear transformation commuting with T.

Theorem 31 is now seen to be a special case of Theorem 29:

Theorem 31. Let T *be a locally nilpotent linear transformation on a vector space. Then a linear transformation commuting with every linear transformation that commutes with* T *is necessarily a power series in* T.

Exercises 82-100

82. Let R be a discrete valuation ring, M the module of type $Z(p^\infty)$ over R. Prove that E(M) is isomorphic to the completion of R. (This shows the inevitability of considering complete valuation rings.)

83. Let R be a principal ideal ring, M an R-module such that E(M) is (ring) isomorphic to R/I, where I is a nonzero ideal in R. Prove that M is (module) isomorphic to R/I.

84. Let R be any ring, M any R-module, e an idempotent in A = E(M). Prove that eAe is isomorphic to the ring of endomorphisms of the module Me.

85. Let R be a discrete valuation ring, M the module of type $Z(p^\infty)$ over R. Prove that an element of E(M) annihilates p^iM if and only if it is divisible by p^i.

86. Let R be a principal ideal ring with two primes p and q such that R/(p) and R/(q) are isomorphic fields. (This is easily arranged by taking R to be a polynomial ring.) Show that the R-modules R/(p) and R/(q) have isomorphic rings of endomorphisms but are not isomorphic as R-modules. (This source of annoyance cannot arise if R is the ring of integers; see the next exercise.)

87. (*a*) Let G and H be (ordinary) primary abelian groups. Show that any isomorphism of their rings of endomorphisms is induced by an isomorphism of G and H.

(*b*) Show that every automorphism of the ring of endomorphisms of a primary abelian group is inner.

(*c*) Extend parts (*a*) and (*b*) to torsion groups.

88. (*a*) Let R be a principal ideal ring, M an R-module annihilated precisely by the nonzero ideal I. Show that E(M) contains an isomorphic copy of R/I.

(*b*) Let N be a second R-module annihilated precisely by I. Show that any (R/I)-isomorphism of E(M) and E(N) is induced by an isomorphism of M and N.

(*c*) Show that the center of E(M) is R/I.

(*d*) Show that any automorphism of E(M) leaving the center elementwise fixed is inner.

89. (*a*) Let R be a complete discrete valuation ring, M and N faithful primary R-modules. Show that any isomorphism of E(M) into E(N) is induced by a semilinear mapping of M onto N. (A mapping V is semilinear if it satisfies $(\alpha x)V = \alpha'(xV)$, where $\alpha \to \alpha'$ is an automorphism of R.)

(*b*) Extend this to nonfaithful modules.

90. Let T be an algebraic linear transformation on a vector space. Show that any linear transformation which commutes with every linear transformation commuting with T is necessarily a polynomial in T. (This is a special case of part (*c*) of exercise 88.)

91. Let R be a principal ideal ring, M a torsion R-module, $M = \Sigma M_p$ its primary decomposition.

(*a*) Prove that E(M) is the *complete* direct sum of the rings $E(M_p)$.

(*b*) Prove that the center of $E(M_p)$ is either R/p^i or R_p^* (in the notation of §14).

(*c*) Prove that the center of E(M) is the complete direct sum of the centers of the rings $E(M_p)$.

92. Let T be a locally algebraic linear transformation on a vector space. Let T_1 be a linear transformation commuting with every linear transformation commuting with T. Prove that on any *finite* number of vectors, T_1 agrees with a suitable polynomial in T. (This is readily deduced from exercise 91. In terms of the standard notion of a weak topology, it may be restated neatly as follows: The double commuting ring of T is the weak closure of the polynomials in T.)

93. Let R be an integral domain with quotient field K, and I a nonzero ideal in R. Let M be the R-module given by $K \oplus R/I$. Show that E(M) is (ring) isomorphic to $K \oplus R/I$ and so is its center. (This example rules out the possibility of any absolutely general theorem asserting that the center of E(M) is R or K. But the conjecture below might stand up: Let R be a complete discrete valuation ring, M a faithful R-module having no divisible torsion-free submodule. *Conjecture*: The center of E(M) is R. See also exercise 95.)

94. Let R be a complete discrete valuation ring, M an R-module. If the group of automorphisms of M is commutative (and, a fortiori, if

E(M) is commutative), show that M is either of rank one, or isomorphic to $K \oplus R/p^i$, where K is the quotient field of R.

95. Let R be any commutative ring with unit, M an R-module having a direct summand isomorphic to R. Show that the center of E(M) is R. (To be proved by a simplification of the technique used in Theorem 29.)

96. Let R be a complete discrete valuation ring. Show that for both the modules R and $Z(p^\infty)$ the ring of endomorphisms is R. (This points up a critical difficulty in extending Theorem 28 to mixed modules. There are reasons for believing that a theory of duality is needed to clarify the situation. Perhaps it is true that when the rings of endomorphisms are isomorphic [or anti-isomorphic], the modules are isomorphic or "dual." Note, however, that this difficulty does not bother us in generalizing Theorem 30. In fact, one may even conjecture that Theorem 30 extends to any module over a complete discrete valuation ring. See the next two exercises.)

97. Let R be a complete discrete valuation ring, M an R-module having R as a direct summand. Prove that every automorphism of A = E(M) leaving the center elementwise fixed is inner. (Let e be the projection for the direct summand. Suppose $e \to f$, $g \to e$ under the isomorphism. If we know that f corresponds to a summand R, the technique of Theorem 28 will work. But otherwise f corresponds to $Z(p^\infty)$, which entails $fAe = 0$, whence $eAg = 0$, which is false no matter whether g corresponds to R or to $Z(p^\infty)$.)

98. Let R be a complete discrete valuation ring, M a torsion-free R-module. Prove that every automorphism of E(M) leaving the center elementwise fixed is inner. (This follows from exercise 97 and Corollary 1 of Theorem 23.)

99. Let R be any ring, M a finitely generated R-module, N the direct sum of a countable number of copies of M. Prove that E(N) is the ring of all row-finite infinite matrices over E(M), i.e., all matrices having only a finite number of nonzero entries in each row.

100. Let R be a complete discrete valuation ring, N the direct sum of a countable number of copies of the module $Z(p^\infty)$ over R. Prove that E(N) is the ring of all infinite matrices over R with the property that each row converges to 0.

20. NOTES

§ 2. A note on terminology: the category-theorists have taught us that we ought to say "direct product" rather than "complete direct sum," but I have maintained the original terminology.

§ 3. No attempt will be made to trace the history of the primary decomposition of torsion groups. The underlying idea certainly goes back to the earliest days of number theory.

Matlis [89], [90] has given a definitive account of the integral domains for which a primary decomposition holds. Let R be an integral domain, M a maximal ideal in R. Call an ideal I an M-ideal if the only maximal ideal containing I is M. Call an R-module A an M-module if the order ideal of any non-zero element of A is an M-ideal. Call R h-local if every prime ideal is contained in a unique maximal ideal and every non-zero ideal is contained in only a finite number of maximal ideals. *Theorem:* R *is* h-*local if and only if every torsion* R-*module is a direct sum of* M-*modules.*

§5. The first general statement of Theorem 2 appeared in Baer's paper [4, p. 766]. He described the result as well known. For countable torsion groups Theorems 2-4 were proved by Zippin [126, pp. 88-90] but his proofs did not really make use of countability.

When §5 was first written, homological algebra was barely visible on the horizon. Although Baer had stated all the basic facts in [7], the enticing name "injective" did not exist. Today we can summarize the facts briefly.

(1) Over any integral domain injective implies divisible.

(2) Over a Dedekind ring, divisible conversely implies injective, as proved in [60]. I did not dare conjecture that this was true only for Dedekind rings, as was proved by Cartan and Eilenberg [17, p. 134, Prop. 5.1].

(3) If R is a (left) Noetherian ring, then any injective (left) R-module is a direct sum of indecomposable ones; this was proved by Matlis [87, Th. 2.5]. The converse was proved by Papp [99] and generalized by Faith and Walker [35].

(4) If R further is commutative, this decomposition is unique up to isomorphism and each summand has the form E(R/P) where P is a prime ideal and E denotes injective envelope [87, Prop. 3.1 and Th. 3.3]. Thus we get a fresh point of view on $Z(p^\infty)$: it is the injective envelope of Z(p). The module E(R/P) is a big brother of $Z(p^\infty)$ and shows a strong family resemblance.

(5) Let R again be arbitrary left Noetherian. Then any left R-module has the form $A \oplus B$ where A is a maximal injective submodule and B has no injective submodules $\neq 0$; A is unique up to isomorphism; if R is left hereditary, A is absolutely unique [87, pp. 512-13].

Over an integral domain that is not Dedekind we still have the companion notion of divisibility to study. A reasonable first question is this: if D is divisible and T is its torsion submodule, is T a direct summand of D? If the quotient field Q of R is countably generated as an R-module, the answer is a quick "yes" (pick a basis of D/T as a vector space over the quotient field, lift these elements in any fashion to D, and build them up to a complement to T by a sequence of suitable divisions by elements of R). But this leaves the uncountable case looking mysterious. Matlis [88] cleared up the mystery nicely: for an affirmative answer we need $d_R(Q) \leq 1$, where $d_R(Q)$ denotes the homological dimension of Q as an R-module. Almost closing the circle of ideas is the theorem [64] that if R has just one maximal ideal then $d_R(Q) \leqq 1$ implies that Q is countably generated.

A final touch from [88]: if a little more is assumed—that D is a homomorphic image of an injective module—then T is a direct summand.

§6. I invented the "test problems" to show that Ulm's theorem could really be used. They have been similarly suggestive in the theory of rings of operators [57]. While they perhaps have no intrinsic importance, I do believe their defeat is convincing evidence that no reasonable invariants exist.

The second test problem was the first to meet defeat, at the hands of Jónsson [55]. In his example the groups G and H were torsion-free of rank two, a possibility foreshadowed in the guide to the literature of the first edition (p. 76). Of course it takes groups of infinite rank to defeat the first test problem, and this was done by Sasiada [110] with uncountable torsion-free groups, then by Corner [23] with countable torsion-free groups. Finally, Crawley [27] defeated both test problems with (of course uncountable) primary groups. Corner and Crawley actually obtained groups G and H related as in exercise 10 of § 6. Subsequently Corner [24] did even better: he has a countable torsion-free group G such that G is isomorphic to $G \oplus G \oplus G$ but not to $G \oplus G$.

Crawley [28] also has interesting results concerning cancellation in primary groups (note that in dealing with countable primary groups, Ulm's theorem allows us to cancel a summand all of whose Ulm invariants are finite).

There was originally a third problem, proposing the cancellation of finitely generated direct summands. It was promptly and ingeniously settled in the affirmative by Cohn [20] and Walker [123]. (It should be noted that the cancellation of a finite direct summand had already been treated by Jónsson and Tarski [56], even in the non-abelian case, and for more general systems than groups.) Hsü [50] pushed on to the case of Dedekind rings.

I take the opportunity to record a generalization of the cancellation theorem for an infinite cyclic direct summand; it is, perhaps, definitive for this circle of ideas.

Theorem. Let R *be a commutative hereditary ring, and* A *and* B R-*modules. Then* $R \oplus A \cong R \oplus B$ *implies* $A \cong B$. *The same is true for semi-hereditary* R, *provided* A *and* B *are finitely generated.*

We may suppose A and B placed in the same module M, with M/A, M/B isomorphic to R. Under either hypothesis (A + B)/B is isomorphic to a projective ideal, hence so is $A/(A \cap B)$. Thus $A \cap B$ is a direct summand of A, and (symmetrically) of B as well. We can therefore suppress this common summand $A \cap B$, and start all over with $R \oplus I \cong R \oplus J$, I and J ideals. If R is a domain, taking exterior product will do the job (as W. Vasconcelos taught me). But the idea behind exterior products can work for general R, and even prove a little more.

Theorem. Let R *be any commutative ring,* I *an ideal in* R, B *an* R-*module. Then* $R \oplus I \cong R \oplus B$ *implies* $I \cong B$.

We phrase the matter as follows: we have a homomorphism f of $R \oplus I$ onto R and wish to prove that the kernel K is isomorphic to I. Suppose that f restricted to R is multiplication by x. We have $ax + f(i) = 1$ for suitable $a \in R$, $i \in I$. We define $g: K \to I$ by sending $(r, s) \in K$ into $ir - as \in I$. We define $h: I \to K$ by $h(z) = (f(z), -xz)$. We have $g(h(z)) = if(z) + axz = z$ since $if(z) = zf(i)$. We have $h(g(r, s)) = (f(ir - as), -x(ir - as)) = (r, s)$ since $xr + f(s) = 0$, $ax + f(i) = 1$, $if(s) = sf(i)$.

We note that the proof still works if I is a fractional ideal (submodule of the total quotient ring of R).

There is a standard application of the ability to cancel R from a direct sum. We note it as a corollary: if in a commutative semi-hereditary ring R the ideal generated by $a_1, \cdots, a_n$ is all of R, then there exists a matrix with this as first row and determinant 1.

§ 7. Suitable extensions of the concept of purity to arbitrary modules over arbitrary rings have received a good deal of attention. We shall devote this note on § 7 to a brief survey.

One might say that in §7 and §8 there were two major results concerning purity: Theorems 5 and 7. A close inspection of the ideas involved in Theorem 5 reveals that one has here a perfect analog of the usual development of projective modules. In a dual fashion, the ideas around Theorem 7 mimic the theory of injective modules; we shall postpone consideration of this dual aspect to the notes on §17.

We begin by copying the definition of purity verbatim. Let R be any ring with unit; we consider left R-modules. A submodule B of A is pure in A if $xB = xA \cap B$ for all x in R.

When we proceed to generalize Theorem 5 we find that the notion of a cyclic module is not exactly right; what we need is a module generated by an element whose annihilator is a principal left ideal, i.e. a module of the form R/Rx. The concept has apparently not yet acquired a name in the literature so we propose the designation cyclically presented (CP for short). Now our generalization of Theorem 5 reads: *if* B *is a pure submodule of* A *and* A/B *is a direct sum of CP modules, then* B *is a direct summand of* A.

The proof of Theorem 5 needs virtually no change, but let us persist in pursuing the parallelism with projectivity. The key lies in the following proposition.

Proposition. B *is pure in* A *if and only if any homomorphism from a CP module* C *to* A/B *can be lifted to* A:

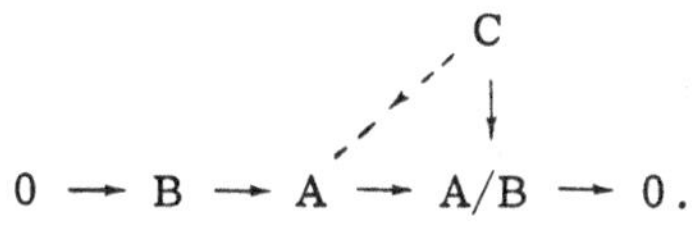

Proof. Suppose B is pure in A. Let C = R/Rx and suppose the image $\bar{1}$ of 1 in C maps to u $\in$ A/B. Pick t in A mapping on u. Then xt $\in$ B $\cap$ xA, so we can write xt = xt_1 with $t_1 \in$ B. Mapping $\bar{1}$ to t - t_1 gives us the required lifted map from C to A.

Suppose the lifting property holds. Let t $\in$ B $\cap$ xA, say t = xa with a $\in$ A. Let u be the image of a in A/B. Then xu = 0 so we can map C to A/B by sending $\bar{1}$ to u. If in the lifted map $\bar{1}$ goes to a_1 then a - $a_1 \in$ B, $xa_1 = 0$, so t = x(a - a_1) $\in$ xB as required.

Call a module *pure-projective* if it shares with CP modules the lifting property in the preceding proposition.

Proposition. A direct sum of modules is pure-projective if and only if each summand is pure-projective.

The proof is the routine one.

Theorem. The following statements are equivalent for a module P:

(1) P *is pure-projective,*

(2) *Whenever* B *is a pure submodule of* A *and* A/B *is isomorphic to* P, B *is a direct summand of* A,

(3) P *is a direct summand of a direct sum of CP modules.*

Proof. (3) → (1). This is immediate from the two propositions.

(1) → (2). By definition there is a mapping from P to A which compounded with the homomorphism from A onto P is the identity on P. This shows B to be a direct summand of A.

(2) → (3). The method is to construct

$$0 \to B \to A \to P \to 0$$

with B pure in A and A a direct sum of CP modules. For A we take a universal object: a direct sum of modules R/Rx, with as many copies of R/Rx as there are homomorphisms of it into P. Then A maps onto P; and that the kernel B is pure checks immediately from the criterion given in the first proposition above.

Let R be an integral domain, I an invertible ideal in R. Then R/I is pure-projective. (Note: this is proved in [60, Lemma 4] when R is Dedekind, but the proof is clumsy and should be replaced by that in [17, p. 133] or [37, p. 251].)

Purity can be recast as follows: it is equivalent to the preservation of exactness when 0 → B → A is tensored with a right CP module R/xR. In his investigation of free products of rings [21], P. M. Cohn found the useful notion to be the requirement that exactness be preserved on tensoring with *any* right module. This version of purity got official recognition from Bourbaki [14]. It may well be the ultimate verdict that Cohn's purity deserves the central role, but we shall call it "strong purity." Logically, but somewhat awkwardly, we call the altered pure-projectivity that goes with it "weak pure-projectivity." Fieldhouse [36] characterized weak pure-projectivity as coinciding with the property of being a direct summand of a direct sum of finitely presented modules. The argument given above is modelled on his.

Warfield [124] proved that the two versions of purity coincide for an integral domain R if and only if R is Prüfer (finitely generated ideals invertible). Fuchs [45], [46] based his study of purity on still another version, which has finitely generated modules playing the role assigned above to cyclically presented or finitely presented modules. Other aspects of purity are studied by Nunke [97] and Stenström [118].

§§ 8-10. Twentieth-century work on infinite abelian groups was inaugurated by Levi [84]. The basic facts (pure subgroups, height, direct sums of cyclic groups) were given by Prüfer (for countable groups) in his ground-breaking paper [102]. The exposition we have given in §§ 7-10 substantially follows Kulikov [73], [74]. It was my delight in reading these two papers, and the availability of Mackey's proof of Ulm's theorem, that led to my teaching of the 1950 course and the subsequent writing of this monograph.

In Prüfer's work one might say that the main result was Theorem 11: A countable primary group with no elements of infinite height is a direct sum of cyclic groups. Naturally, this covers Theorem 6 as a special case: A group of bounded order is a direct sum of cyclic groups. It did not appear clearly until later that the first of these results is false for uncountable groups, while the second is true.

There is another possible point of view concerning groups of bounded order—we can regard them as modules over the ring of integers mod n. Or, to make the appropriate generalization, we may study modules over a principal ideal ring with the descending-chain condition. It turns out that they are direct sums of cyclic modules; indeed, the proof of Theorem 6 goes through with only nominal changes. It is in this spirit that the problem is treated by Köthe [71] and Asano [1]; they, moreover, allow the ring to be noncommutative. In the commutative case, Köthe proved a converse theorem: If R is a commutative ring with unit element and descending-chain condition, such that every R-module is a direct sum of cyclic modules, then R is a principal ideal ring. Cohen and Kaplansky [19] perfected Köthe's theorem by showing that the assumption of the descending-chain condition is redundant. Chase [18] and Faith and Walker [35] derived the descending-chain condition from weaker hypotheses.

Since there has been some correspondence concerning exercise 29 in § 10 I note its appearance in Fuchs [39, Th. 24.8 on p. 81], attributed there to M. Erdelyi [33].

The late Tibor Szele in his review (*Mathematical Reviews* 16 [1955], 444-46) scolded me for ignoring basic subgroups. I confess that I thought exercise 22 in §9 was sufficient recognition. I was wrong; at the hands of numerous experts they have proved extremely useful. They appear in the new version of § 16.

§11. Ulm's proof [119] used rather complicated manipulations of infinite matrices. Zippin [126] gave a direct group-theoretic proof, as did Kurosh [78], [79]. A discerning, thorough account appears in Fuchs [39, pp. 117-36].

As a prelude to work on operators on Banach spaces, George Mackey made a purely algebraic study of locally algebraic linear transformations on an $\aleph_0$-dimensional vector space. In doing so, he rediscovered Ulm's theorem. His proof made its first appearance as a series of exercises in Bourbaki [12, pp. 79-81], then in a version which admitted a torsion-free part of rank one [65], and subsequently in *Infinite Abelian Groups*. In [105], [109], [93], and [94] the ideas of [65] are developed further.

The existence of groups with prescribed Ulm invariants is treated by Ulm, Zippin, Kurosh, and Fuchs. The method sketched in exercises 40-42 of §11 was adapted by me from the unpublished treatise by Baer mentioned in the Introduction. More general existence theorems are proved by Kulikov in [75].

Baer [2] made an interesting application of Ulm's theorem. He showed that if G is a reduced countable primary group which is not a direct sum of cyclic groups (i.e. the length of G exceeds ω) then G always admits two direct sum decompositions which do not have isomorphic refinements. What if G is uncountable? Even if G has elements of infinite height, the refinement theorem may hold, as noted by Corner and Crawley in [26]. When there are no elements of infinite height, Crawley gave certain sufficient conditions for refinement to hold in [29], but this is not always true [26].

The exposition in §11 put us in possession of two classes of reduced primary groups for which the Ulm invariants suffice: direct sums of cyclic groups and countable groups. Kolettis [67] took the step of unifying these by studying direct sums of countable groups; he proved that again the Ulm invariants suffice. Simplified accounts were presented by Hill [52], Crawley [30], and Richman and Walker [104]. Nunke [98] introduced a significantly wider class of groups, which he called *totally projective*. Hill [53] has announced that Ulm's theorem extends to these groups. In [31] Crawley and Hales prove Ulm's theorem for groups admitting a representation with relations of length at most two; using [53], they identify their groups with Nunke's. Related ideas are studied by Parker and Walker in [100].

I conclude with a note concerning exercise 33. Kolettis [68] amplified this example by studying a direct sum $G \oplus H$ where G is a direct sum of cyclic groups and H is the torsion subgroup of a complete direct sum of cyclic groups. In this way one quickly exhibits continuum many non-isomorphic groups all having Ulm invariants 1, 1, 1, $\cdots$. By deeper methods Leptin [83] found the precise number to be 2^c (c = the cardinal of the continuum). Peter Crawley has informed me that by the methods of [29, §4] he can prove the following:

Let $a_0, a_1, a_2, \cdots$ be a sequence of cardinals such that, for each i, the sum from the i-th term on is an infinite cardinal n not exceeding 2^c. Then the number of non-isomorphic p-groups with no elements of infinite height having Ulm invariants $a_0, a_1, a_2, \cdots$ is precisely 2^{c+n}. He conjectures that the restriction $n \leq 2^c$ is not needed.

§15. The notes on this section will be divided under a number of subheadings.

Finitely generated modules. Probably the best known proof of Theorem 16 uses matrix methods which proceed by diagonalizing the matrix of relations satisfied by a set of generators. These methods are analyzed closely in [59]. This is a good place to record what ought to have been the main theorem in [59] but was never clearly stated: for an integral domain R the following statements are equivalent, (1) R is an elementary divisor ring, (2) every finitely presented R-module is a direct sum of cyclic modules.

We observe in passing that the matrix methods prove a little more, namely the possibility of choosing "simultaneous bases" for a finitely generated free module and a submodule. A generalization to the infinite case appears in exercise 80(a) of §18, but the more satisfactory generalization mentioned in part (b) remains in doubt.

The following is still an open question: is every Bézout domain (i.e. domain in which finitely generated ideals are principal) an elementary divisor ring? To the results in [59] the following unpublished ones concerning a Bézout domain R may be added: if R has a countable number of maximal ideals, or if non-zero prime ideals in R are maximal, then R is an elementary divisor ring (in the second case it is even "adequate").

Uzkov [121] proved that if R is commutative Noetherian (divisors of zero allowed), and every finitely generated R-module is a direct sum of cyclic modules, then R is a principal ideal ring.

We return to the consideration of non-Noetherian domains. There is a pertinent additional class for which it is true that finitely generated modules are direct sums of cyclic modules: the almost-maximal valuation rings introduced in [60]. Today we can give a swift definition to replace the awkward systems of congruences: assume K/R injective, where K is the quotient field of the valuation ring R.

Suppose a domain R has the property that finitely generated R-modules are direct sums of cyclic modules. As far as known examples go, it is a tenable conjecture that R must be a principal ideal ring or an almost-maximal valuation ring. Matlis [90, Th. 5.7] has made an advance toward the goal; assuming that R is h-local (see the notes on §3) he proves that R_M is an almost-maximal valuation ring for every maximal ideal M in R.

Splitting mixed groups. The pioneering result, due independently and about simultaneously to Fomin [38] and Baer [4], is that the torsion subgroup is a direct summand if it is of bounded order. Baer proved at once that this result is definitive: if a torsion group has this universal splitting property then it is the direct sum of a divisible group and a group of bounded order.

Baer also posed the dual question: if a torsion-free group G has the property that a group splits whenever the torsion-free factor group is G, must G be free? He proved it when G is countable and made some headway when G is uncountable.

Twenty years later Baer [10] returned to the problem and made further progress: he proved that an infinite complete direct sum of cyclic groups does not have this universal splitting property (a problem specially noted in the first edition). J. Erdős [34] did the same thing at almost the same time. According to Fuchs [39, p. 190], Sasiada had an unpublished proof also at about the same time.

Ten years later still, Griffith [47] solved the problem completely: for any cardinal number such a group must be free. The principal point in the proof is of independent interest: he constructs a group H with torsion subgroup T such that (1) H/T is divisible and arbitrarily large, (2) every torsion-free subgroup of H is free.

When the framework is broadened to arbitrary integral domains, there are of course new problems. For instance, it is tricky to prove that the splitting of every module implies that the domain is a field [106]. Chase [18, Th. 4.3] proves: if a domain R has the property that the torsion submodule is always a direct summand when it is of bounded order, then R is Dedekind. In [63] some general steps are taken in studying the problem over an arbitrary integral domain.

Torsion-free groups. In this strange part of the subject anything that can conceivably happen actually does happen. The existence of indecomposable groups up to the first strongly inaccessible cardinal is shown by Corner [25], correcting and completing Fuchs [42]. Jónsson [55] has three minimal examples of bad behavior in torsion-free groups of finite rank: a counter-example to Test Problem II with G and H of rank two, a group of rank 3 that decomposes in two different ways, and a group rank 4 that can be split into indecomposables as 1 + 3 or as 2 + 2. Bourbaki [13, Ex. 11, p. 138] has assembled these examples in capsule form. Corner [22] shows that a torsion-free group of finite rank can exhibit all possible combinations into a fixed number of indecomposable summands. In [23] he proves that every countable reduced torsion-free ring is an endomorphism ring, a remarkable theorem with many remarkable corollaries.

The invariants of Kurosh [77], Malcev [85], and Derry [32] for torsion-free groups of finite rank lie half forgotten. It is a fair assessment by Fuchs [39, p. 157] that they translate the given problem into another that is at least as hard.

Concerning Theorem 19. Suppose R is a principal ideal ring which has no indecomposable torsion-free module of rank 2. Let p be a prime in R. The hypothesis is clearly inherited by R_p. So by Theorem 19 every R_p is complete.

In the first edition it was concluded, without any justification, that R could only have one prime. I returned to this in [62] but again there was an error. (See also [91].) Two swings and two strikes!

However, a famous theorem of F. K. Schmidt [112] on multiply complete fields is applicable, and settles the question. My having overlooked this is all the more incredible since Schmidt's theorem was generalized in a paper by myself and O. F. G. Schilling [66].

Mrs. Barbara Osofsky gave a simple direct argument (transmitted by E. Matlis). Suppose R_p is complete and q is a different prime. Solve $x \equiv q \pmod{q^2}$, $x \equiv 1 \pmod{p}$. Then by the Hensel lemma x is a k-th power for k prime to the characteristic of R/(p), impossible since x is exactly divisible by q.

Concerning Theorem 20. This theorem generalizes to maximal valuation rings [60]. (We can define a maximal valuation ring as an almost-maximal one which is furthermore complete in the natural topology of the ring, or as a linearly compact valuation ring, or we can go back to systems of congruences, or back to pseudo-convergent sequences, or all the way back to the original impossibility of extending the ring without enlarging the value group or residue class field.)

Fleischer [37] added the theorem that in a torsion-free module over a maximal valuation ring, every pure submodule of finite rank is a direct summand.

Let a domain R have the property that all its torsion-free modules of finite rank are direct sums of modules of rank one. What are the possibilities for R? The full answer is not known, but Matlis [91] has substantial partial results. In particular he has found the exact answer in the Noetherian case: complete local domains of Krull dimension one in which every ideal can be generated by two elements.

Concerning Theorem 21. In the early 1940's I heard a number of mathematicians asking whether a complete direct sum of infinite cyclic groups is free. I put the question to Paul Erdős. He promptly came up with a proof that worked for any principal ideal ring with an infinite number of primes. In this form it appeared in Bourbaki [12, p. 84, Ex. 10]. All this time a much better proof by Baer was sitting unobserved [6, p. 121]. This proof is reproduced in the text.

Krull [72, Th. 16] subsequently proved: if R is a domain such that a complete direct sum of an infinite number of copies of R is free then R is a field. Chase [18] improved this by replacing "free" by "projective." Chase's paper should be consulted for many related interesting results.

Direct sums of torsion-free modules of rank one. Is this property inherited by direct summands? Baer proposed this problem in [6] and answered it affirmatively subject to a certain chain condition. In the first edition I described the problem as "particularly challenging." It is pleasant to report that the challenge was met. Kulikov [76] removed the chain condition but needed a countability hypothesis; Fuchs [40] simplified the proof. The subject is full of surprises; in this case bridging the countability gap was the easiest part of the job [61].

Kolettis [70] extended the theorem to Dedekind rings.

Modules of finite-valued functions. A theorem proved by Specker in [117] remains one of the most remarkable in the subject: in the group of all bounded sequences of integers all subgroups of cardinal $\aleph_1$ are free (and hence the continuum hypothesis implies that the group itself is free).

Let us note how the word "bounded" should be reinterpreted when we switch to general rings: we should speak about sequences (or functions) taking on only a finite number of values. Let then R be a principal ideal ring, A the module of all finite-valued functions on a countable set. Then any submodule of A generated by $\aleph_1$ elements is free. Specker's proof needs only nominal changes.

Here is a curious remark: if $2R = 0$ we can prove that A is free without the continuum hypothesis. In fact R can be absolutely any ring with $2R = 0$. The proof is to look at the idempotents (characteristic functions) in A as a vector space over the field of two elements and take a basis; this is also a basis for A.

§16. As was acknowledged in the introduction (§1), the torsion-free case of Theorem 22 is attributable to Isidore Fleischer. The companion investigation for torsion modules was made by Kulikov [74] and Kaloujnine [58]. Exercise 58 is to be found in both of these papers.

It was on examining these two parallel theories that it occurred to me to combine them in a general theory of complete modules. The rewritten version has benefitted greatly from the sprightly paper [107] of Rotman.

§17. The study of algebraic compactness made in the first edition of *Infinite Abelian Groups* had its origin back in 1944 when I read Halmos' paper [49]. Years later, with the theory of complete modules at hand, I realized I was in a position to tell my colleague Paul Halmos exactly what were the possibilities for the algebraic structure of a compact abelian group (modulo some final fiddling with cardinal numbers). It was in order to be able to state a precise theorem without this annoying investigation of cardinal numbers that I broadened the class of compact groups slightly and called the generalized object "algebraically compact." There was a lucky accident here, for the concept had ramifications considerably beyond anything I envisaged.

The definition of an algebraically compact (AC) module in the text is meaningful for any ring as soon as we agree on purity. We shall (except for a remark toward the end) take purity to be defined as in the notes on § 7. It is apparent that the concept is dual to the pure-projectivity that we discussed there. The situation calls for the following proposition.

Proposition. Let R *be any ring,* Q *a (left)* R-*module. The following statements are equivalent:*

(1) Q *is a direct summand of any module containing it as a pure submodule (i.e.* Q *is* AC),

(2) *Given modules* $B \subset A$ *with* B *pure, and a homomorphism* f *from* B *to* Q, f *can be extended to a homomorphism from* A *to* Q. *In diagram form:*

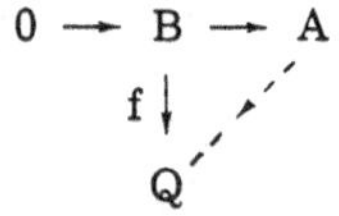

That (2) implies (1) is obvious. In proving that (1) implies (2) one might be tempted to imitate the projective procedure; this would call for a pure embedding of a general module in an AC one (analogous to embedding a module in an injective one). It is perhaps not yet universally known that the embedding can be avoided by use of a suitable pushout (and the analogous argument is equally feasible in the projective case). Sketch: let $D = A \oplus Q$, let E be the submodule of D consisting of all $\{b, f(b)\}$, and let $G = D/E$. Then we verify that the natural map from Q to G is one-to-one, that (after identification) this makes Q a pure submodule of G, and that the fact that Q is a direct summand of G gives us (via the induced retraction of G on Q) the extension of f to A which we seek.

Remarks. 1. This proposition trivializes the proof of exercises 63 and 64.

2. The designation "pure-injective" is highly appropriate, and it might be a good idea to have it permanently replace "algebraically compact."

3. Over a Dedekind ring the rest of the discussion we gave in the notes on § 7 can be dualized. Purity can be characterized by mappings into torsion cyclic modules and $Z(p^\infty)$'s (and these may be neatly combined as "cocyclic" modules); a module is AC if and only if it is a direct summand of a complete direct sum of cocyclic modules. With some additional touches, this can lead up to a proof of the result in exercise 65. As references we mention the pioneering paper of Maranda [86], the accounts by Fuchs [39, pp. 83-87], [44, pp. 10-12] and the references given there, and the papers [36], [45], [46], [118], [124].

4. Corresponding to Cohn's strengthened purity there is a weakened notion of AC to be studied.

5. I find it noteworthy that AC modules over a Dedekind ring have a "primary decomposition" of just the kind hoped for in §14. This is in line with the method so widely used in the theory of commutative Noetherian rings: localize, complete, use structure theory, then (hopefully) work your way back. At any rate, I advise new workers in the field: consider trying your problems one prime at a time. (Parenthetical note: [69] is an intriguing (perhaps unique) case where something works with two primes but not with more than two.)

6. Since the structure of a compact abelian group played such a big role in the first edition, and since it was treated a little inadequately, we shall reprise it here. Fuchs made an excellent suggestion in [41]: we should borrow from Pontrjagin the fact that G = Hom (A, D), D the reals mod one, and then proceed purely algebraically.

So let R be a Dedekind ring, A an R-module, D a divisible R-module, C = Hom (A, D). Let T be the torsion submodule of A. Then Hom (A/T, D) is a divisible module, in a natural way a submodule of C, thus a direct summand of C. We might as well start over with A torsion, and then D might as well be torsion. Primary decomposition works: after allowing for a complete direct sum we can assume that R is a complete discrete valuation ring and that A and D are primary.

I shall finish the analysis assuming $D = Z(p^\infty)$, as would be the case if we had started with the reals mod one. Let B be a basic submodule of A. Then Hom (A/B, D) is a complete direct sum of copies of R, it is a pure submodule of C, and hence a direct summand. The other summand, Hom (B, D), is a complete direct sum of $Z(p^n)$'s.

This complete direct sum decomposition, due to Fuchs [41], Hulanicki [54], and Harrison [51], is a nice improvement over my original description of the primary parts as simply being complete. It trivializes the job of specifying what cardinal numbers are eligible.

In conclusion I shall mention the groups felicitously named "cotorsion" by Harrison [51]. The same class of groups was discovered independently, and almost simultaneously, by Fuchs [43] who called them "B-groups," and Nunke [96] who called them "realizable." We have a hierarchy of three successively weaker demands placed on a module A: always a direct summand (divisible), a direct summand when it is pure (AC), a direct summand when the quotient is torsion-free (cotorsion). Cotorsion modules over a Dedekind ring have a pretty structure theory: after we discard divisible and torsion-free direct summands, the surviving object is determined by its torsion submodule, and this can be any torsion module.

Cotorsion modules over an arbitrary integral domain were studied by Matlis [89].

§18. This section is really a research paper on a rather special subject. It got into the monograph because I was excited about it at the time. The material descends directly from the work of Shoda [115], Baer [3], and Shiffman [114]. These authors considered primary groups with increasing generality: finite (Shoda), direct sum of cyclic groups (Baer), no elements of infinite height (Shiffman); Baer also allowed a divisible subgroup. Some additional considerations were needed here to take care of the complications of transfinite height. The fact that the lattice of fully invariant submodules satisfies only one of the two infinite distributive laws still intrigues me.

Beaumont and Pierce carried the investigation further in [11].

In the first edition I optimistically wrote "It seems plausible to conjecture that every R-module is both transitive and fully transitive." This conjecture has been defeated by Megibben [92], and so we have another striking example of the sharp difference between the countable and uncountable in the subject.

Remark. The constant reference to complete discrete valuation rings in §18 has proved somewhat misleading to readers. Except in exercise 72, this is pure propaganda. Substitute "principal ideal ring" at will.

§19. Theorem 28 was inspired by Baer [9]. In fact, Baer proved Theorem 28 for modules of bounded order having at least three independent elements of maximum order. However, the techniques of the proofs are quite different. The proof of Theorem 28 is ring-theoretic and reminiscent of manipulations with matrices; Baer's is based on his extensive theory of the lattice of submodules.

This theory deserves comment on its own. One way to motivate it is to recall that a projective geometry is nothing but the lattice of subspaces of a vector space over a division ring. We can generalize, in a way that also covers finite abelian groups, if we replace the division ring by a suitable kind of principal ideal ring. (Among other things, it is highly desirable that we allow the principal ideal ring to be noncommutative, for we do not wish to assume Pappus' theorem before even getting started.) This program was carried out in full by Baer [8]: geometric axioms, introduction of coordinates, uniqueness and mapping theorems.

The application of Theorem 29 to Theorem 31 was suggested by the finite-dimensional proof of the double commutator given by Lagerstrom [80].

Studies of the ring of endomorphisms have been augmented by more difficult ones on the group of automorphisms, e.g. [82].

Results in the torsion-free case have been obtained by Wolfson [125].

I am informed by Charles Megibben that the conjecture in exercise 93 is correct.

BIBLIOGRAPHY

Useful introductions are to be found in [48, Chs. 3 and 13], [108, Ch. 9], [111, Ch. 2], [113, Ch. 5], and [116, Ch. 2.4].

Where a reference is incomplete, the paper had not yet appeared at the time the bibliography was drawn up.

1. K. Asano, Über verallgemeinerte abelsche Gruppen mit hyperkomplexem Operatorenring and ihre Anwendungen, Jap. J. Math. 15 (1939), 231-53.

2. R. Baer, The decomposition of abelian groups into direct summands, Quart. J. of Math. 6 (1935), 217-21.

3. ——— Types of elements and the characteristic subgroups of abelian groups, Proc. Lon. Math. Soc. 39 (1935), 481-514.

4. ——— The subgroup of elements of finite order of an abelian group, Ann. of Math. 37 (1936), 766-81.

5. ——— Primary abelian groups and their automorphisms, Amer. J. of Math. 59 (1937), 99-117.

6. ——— Abelian groups without elements of finite order, Duke Math. J. 3 (1937), 68-122.

7. ——— Abelian groups that are direct summands of every containing abelian group, Bull. Amer. Math. Soc. 46 (1940), 800-806.

8. ——— A unified theory of projective spaces and finite abelian groups, Trans. Amer. Math. Soc. 52 (1942), 283-343.

9. ——— Automorphism rings of primary abelian operator groups, Ann. of Math. 44 (1943), 192-227.

10. ——— Die Torsionsuntergruppe einer abelschen Gruppe, Math. Ann. 135 (1958), 219-34.

11. R. A. Beaumont and R. S. Pierce, Partly invariant submodules of a torsion module, Trans. Amer. Math. Soc. 91 (1959), 220-30.

12. N. Bourbaki, Algèbre Ch. VII, Modules sur les anneaux principaux, first edition, Paris, 1952.

13. ——— ibid, second edition, Paris, 1964.

14. ——— Algèbre Commutative, Ch. I, § 2, ex. 24.

15. J. Braconnier, Sur les groupes topologiques localements compacts, J. Math. Pures Appl. 27 (1948), 1-85.

16. A. Brown, The unitary equivalence of binormal operators, Amer. J. of Math. 76 (1954), 414-34.

17. H. Cartan and S. Eilenberg, Homological Algebra, Princeton, 1956.

18. S. Chase, Direct products of modules, Trans. Amer. Math. Soc. 97 (1960), 457-73.

19. I. S. Cohen and I. Kaplansky, Rings for which every module is a direct sum of cyclic modules, Math. Zeit. 54 (1951), 97-101.

20. P. M. Cohn, The complement of a finitely generated direct summand of an abelian group, Proc. Amer. Math. Soc. 7 (1956), 520-21.

21. ——— On the free product of associative rings, Math. Zeit. 71 (1959), 380-98.

22. A. L. S. Corner, A note on rank and direct decompositions of torsion-free abelian groups, Proc. Camb. Phil. Soc. 57 (1961), 230-33.

23. ——— Every countable reduced torsion-free ring is an endomorphism ring, Proc. Lon. Math. Soc. 13 (1963), 687-710.

24. ——— On a conjecture of Pierce concerning direct decompositions of abelian groups, Proc. of Coll. on Abelian Groups, Budapest, 1964, 43-48.

25. ——— Endomorphism algebras of large modules with distinguished submodules.

26. A. L. S. Corner and P. Crawley, An abelian p-group without the isomorphic refinement property.

27. P. Crawley, Solution of Kaplansky's test problems for primary abelian groups, J. Alg. 2 (1965), 413-31.

28. ——— The cancellation of torsion abelian groups in direct sums, J. Alg. 2 (1965), 432-42.

29. ——— An isomorphic refinement theorem for certain abelian p-groups, J. of Alg. 6 (1967), 376-87.

30. ——— Abelian p-groups determined by their Ulm sequences, Pac. J. of Math. 22 (1967), 235-39.

31. P. Crawley and A. W. Hales, The structure of torsion abelian groups given by presentations.

32. D. Derry, Über eine Klasse von abelschen Gruppen, Proc. Lon. Math. Soc. 43 (1937), 490-506.

33. M. Erdelyi, Direct summands of abelian torsion groups, Acta. Univ. Debrecen 2 (1955), 145-49 (Hungarian).

34. J. Erdős, On the splitting problem of mixed abelian groups, Pabl. Math. Debrecen 5 (1958), 364-77.

35. C. Faith and E. A. Walker, Direct-sum representations of injective modules, J. of Alg. 5 (1967), 203-21.

36. D. Fieldhouse, Purity and flat covers, Queen's University preprint no. 14, 1967.

37. I. Fleischer, Modules of finite rank over Prüfer rings, Ann. of Math. 65 (1957), 250-54.

38. S. Fomin, Über periodische Untergruppen der unendlichen abelschen Gruppen, Mat. Sbornik 2 (1937), 1007-9.

39. L. Fuchs, Abelian Groups, Budapest, 1958.

40. ——— Notes on abelian groups. I. Ann. Univ. Sci. Budapest Eötvös Sect. Math. 2 (1959), 5-23.

41. ——— On character groups of discrete abelian groups, Acta Math. Acad. Sci. Hungar. 10 (1959), 133-40.

42. ——— The existence of indecomposable abelian groups of arbitrary power, Acta Math. Acad. Sci. Hungar. 10 (1959), 453-57.

43. ——— Notes on abelian groups. II. Acta Math. Acad. Sci. Hungar. 11 (1960), 117-25.

44. ——— Recent results and problems on abelian groups, in Topics in Abelian Groups, Scott, Foresman and Co., 1963, 9-40.

45. ——— Algebraically compact modules over Noetherian rings, Ind. J. of Math.

46. ——— Note on purity and algebraic compactness for modules.

47. P. Griffith, A solution to the splitting mixed group problem of Baer.

48. M. Hall, The Theory of Groups, Macmillan, New York, 1959.

49. P. Halmos, Comment on the real line, Bull. Amer. Math. Soc. 50 (1944), 877-78.

50. C.-S. Hsü, Theorems on direct sums of modules, Proc. Amer. Math. Soc. 13 (1962), 540-42.

51. D. K. Harrison, Infinite abelian groups and homological methods, Ann. of Math. 69 (1959), 366-91.

52. P. Hill, Sums of countable primary groups, Proc. Amer. Math. Soc. 17 (1966), 1469-70.

53. ——— Ulm's theorem for totally projective groups.

54. A. Hulanicki, Algebraic structure of compact abelian groups, Bull. Acad. Pol. Sci. 6 (1958), 71-73.

55. B. Jónsson, On direct decomposition and torsion-free abelian groups, Math. Scand. 5 (1957), 230-35.

56. B. Jónsson and A. Tarski, Direct decomposition of finite algebraic systems, Univ. of Notre Dame, 1947.

57. R. V. Kadison and I. M. Singer, Three test problems in operator theory, Pac. J. of Math. 7 (1957), 1101-6.

58. L. Kaloujnine, Sur les groupes abéliens primaires sans éléments de hauteur infinie, C. R. Acad. Sci. Paris 225 (1947), 713-15.

59. I. Kaplansky, Elementary divisors and modules, Trans. Amer. Math. Soc. 66 (1949), 464-91.

60. ——— Modules over Dedekind rings and valuation rings, Trans. Amer. Math. Soc. 72 (1952), 327-40.

61. ——— Projective modules, Ann. of Math. 68 (1958), 372-77.

62. ——— Decomposability of modules, Proc. Amer. Math. Soc. 13 (1962), 532-35.

63. ——— The splitting of modules over integral domains, Arch. Math. 13 (1962), 341-43.

64. ——— The homological dimension of a quotient field, Nagoya Math. J. 27 (1966), 139-42.

65. I. Kaplansky and G. W. Mackey, A generalization of Ulm's theorem, Summa Brasil Math. 2 (1951), 195-202.

66. I. Kaplansky and O. F. G. Schilling, Some remarks on relatively complete fields, Bull. Amer. Math. Soc. 48 (1942), 744-47.

67. G. Kolettis, Direct sums of countable groups, Duke Math. J. 27 (1960), 111-25.

68. ——— Semi-complete primary abelian groups, Proc. Amer. Math. Soc. 11 (1960), 200-205.

69. ——— A theorem on pure submodules, Can. J. Math. 12 (1960), 483-87.

70. ——— Homogeneously decomposable modules.

71. G. Köthe, Verallgemeinerte abelsche Gruppen mit hyperkomplexem Operatorenring, Math. Zeit. 39 (1935), 31-44.

72. W. Krull, Über geordnete Gruppen von reelen Funktionen, Math. Zeit. 64 (1956), 10-40.

73. L. Kulikov, Zur Theorie der abelschen Gruppen von beliebiger Mächtigheit, Mat. Sbornik 9 (1941), 165-81. (Russian with German summary.)

74. ——— On the theory of abelian groups of arbitrary power, Mat. Sbornik 16 (1945), 129-62. (Russian with English summary.)

75. L. Kulikov, Generalized primary groups I. Trudy Moskov Mat. Obsc 1 (1952), 247-326. II - Same Trudy 2 (1953), 85-167. (Russian.)

76. ——— On direct decompositions of groups, Ukrain Mat. Z 4 (1952), 230-75, 347-72. (Russian.) AMS Translation, 1956.

77. A. Kurosh, Primitive torsionfreie abelsche Gruppen von endlichen Range, Ann. of Math. 38 (1937), 175-203.

78. ——— Theory of Groups, Moscow, 1944. (Russian.) German translation, Berlin, 1953.

79. ——— Theory of groups, second edition, Moscow, 1953 (Russian). English translation, Chelsea, New York, 1955-56.

80. P. Lagerstrom, A proof of a theorem on commutative matrices, Bull. Amer. Math. Soc. 51 (1945), 535-36.

81. S. Lefschetz, Algebraic topology, AMS Coll. Publ. no. 27, New York, 1942.

82. H. Leptin, Abelsche p-Gruppen und ihre Automorphismengruppen, Math. Zeit 73 (1960), 235-53.

83. ——— Zur Theorie der überabzählbaren abelschen p-Gruppen, Abhandl. Math. Sem. Univ. Hamb. 24 (1960), 79-90.

84. F. W. Levi, Abelsche Gruppen mit abzählbaren Elementen, Leipzig, 1917.

85. A. I. Malcev, Torsionfreie abelsche Gruppen vom endlichen Rang, Mat. Sbornik 4 (1938), 46-58. (Russian with German summary.)

86. J.-M. Maranda, On pure subgroups of abelian groups, Arch. der Math. 11 (1960), 1-13.

87. E. Matlis, Injective modules over Noetherian rings, Pac. J. Math. 8 (1958), 511-28.

88. ——— Divisible modules, Proc. Amer. Math. Soc. 11 (1960), 385-91.

89. ——— Cotorsion modules, Mem. Amer. Math. Soc. no. 49, 1964.

90. ——— Decomposable modules, Trans. Amer. Math. Soc. 125 (1966), 147-79.

91. ——— The decomposability of torsion-free modules of finite rank, Trans. Amer. Math. Soc.

92. C. Megibben, Large subgroups and small homomorphisms, Mich. Math. J. 13 (1966), 153-60.

93. ——— On mixed groups of torsion-free rank one, Ill. J. of Math. 11 (1967), 134-44.

94. C. Megibben, Modules over an incomplete discrete valuation ring, Proc. Amer. Math. Soc. 19 (1968), 450-52.

95. T. Motzkin, The Euclidean algorithm, Bull. Amer. Math. Soc. 55 (1949), 1142-46.

96. R. Nunke, Modules of extensions over Dedekind rings, Ill. J. of Math. 3 (1959), 222-41.

97. ——— Purity and subfunctors of the identity, in Topics in Abelian Groups, Scott, Foresman and Co. 1963, 121-71.

98. ——— Homology and direct sums of countable abelian groups, Math. Zeit. 101 (1967), 182-212.

99. Z. Papp, On algebraically closed modules, Publ. Math. Debrecen 6 (1959), 311-27.

100. L. D. Parker and E. A. Walker, An extension of the Ulm-Kolettis theorems.

101. R. S. Pierce, Homomorphisms of primary abelian groups, in Topics in Abelian Groups, Scott, Foresman and Co. 1963, 215-310.

102. H. Prüfer, Untersuchungen über die Zerlegbarkeit der abzählbaren primären abelschen Gruppen, Math. Zeit. 17 (1923), 35-61.

103. ——— Theorie der abelschen Gruppen, I, Grundeigenschaften, Math. Zeit. 20 (1924), 165-87; II, Ideale Gruppen, Math. Zeit. 22 (1925), 222-49.

104. F. Richman and E. A. Walker, Extending Ulm's theorem without group theory.

105. J. Rotman, Mixed modules over valuation rings, Pac. J. of Math. 10 (1960), 607-23.

106. ——— A characterization of fields among integral domains, An. Acad. Bras. Ci. 32 (1960), 193-94.

107. ——— A note on completions of modules, Proc. Amer. Math. Soc. 11 (1960), 356-60.

108. ——— The Theory of Groups: An Introduction, Allyn and Bacon, Boston, 1965.

109. J. Rotman and T. Yen, Modules over a complete discrete valuation ring, Trans. Amer. Math. Soc. 98 (1961), 242-54.

110. E. Sasiada, Negative solution of I. Kaplansky's first test problem for abelian groups and a problem of K. Borsuk concerning cohomology groups, Bull. Acad. Polon. 9 (1961), 331-34.

111. E. V. Schenkman, Group Theory, Van Nostrand and Co., Princeton, 1965.

112. F. K. Schmidt, Mehrfach perfekte Körper, Math. Ann. 108 (1933), 1-25.

113. W. R. Scott, Group Theory, Prentice-Hall, 1964.

114. Max Shiffman, The ring of automorphisms of an abelian group, Duke Math. J. 6 (1940), 579-97.

115. K. Shoda, Über die characteristichen Untergruppen einer endlichen abelschen Gruppe, Math. Zeit. 31 (1930), 611-24.

116. W. Specht, Gruppentheorie, Springer-Verlag, 1956.

117. E. Specker, Additive Gruppen von Folgen ganzer Zahlen, Portugaliae Math. 9 (1950), 131-40.

118. B. T. Stenström, Pure submodules, Arkiv för Mat. 7 (1967), 159-71.

119. H. Ulm, Zur Theorie der abzählbar—unendlichen abelschen Gruppen, Math. Ann. 107 (1933), 774-803.

120. ——— Elementarteilertheorie unendlicher Matrixen, Math. Ann. 114 (1937), 493-505.

121. A. I. Uzkov, On the decomposition of modules over a commutative ring into direct sums of cyclic submodules, Mat. Sbornik 62 (1963), 469-75. (Russian.)

122. N. Vilenkin, Direct decompositions of topological groups, I. II Mat. Sbornik 19 (1946), 85-154, 311-40. (Russian.) AMS Translation no. 23.

123. E. A. Walker, Cancellation in direct sums of groups, Proc. Amer. Math. Soc. 7 (1956), 898-902.

124. R. B. Warfield, Jr., Purity and algebraic compactness for modules, Pac. J. of Math.

125. K. G. Wolfson, Isomorphisms of the endomorphism rings of torsion-free modules, Proc. Amer. Math. Soc. 13 (1962), 712-14.

126. L. Zippin, Countable torsion groups, Ann. of Math. 36 (1935), 86-99.

INDEX

A CATALOG OF SELECTED

DOVER BOOKS

IN SCIENCE AND MATHEMATICS

Mathematics–Algebra and Calculus

VECTOR CALCULUS, Peter Baxandall and Hans Liebeck. This introductory text offers a rigorous, comprehensive treatment. Classical theorems of vector calculus are amply illustrated with figures, worked examples, physical applications, and exercises with hints and answers. 1986 edition. 560pp. 5 3/8 x 8 1/2. 0-486-46620-5

ADVANCED CALCULUS: An Introduction to Classical Analysis, Louis Brand. A course in analysis that focuses on the functions of a real variable, this text introduces the basic concepts in their simplest setting and illustrates its teachings with numerous examples, theorems, and proofs. 1955 edition. 592pp. 5 3/8 x 8 1/2. 0-486-44548-8

ADVANCED CALCULUS, Avner Friedman. Intended for students who have already completed a one-year course in elementary calculus, this two-part treatment advances from functions of one variable to those of several variables. Solutions. 1971 edition. 432pp. 5 3/8 x 8 1/2. 0-486-45795-8

METHODS OF MATHEMATICS APPLIED TO CALCULUS, PROBABILITY, AND STATISTICS, Richard W. Hamming. This 4-part treatment begins with algebra and analytic geometry and proceeds to an exploration of the calculus of algebraic functions and transcendental functions and applications. 1985 edition. Includes 310 figures and 18 tables. 880pp. 6 1/2 x 9 1/4. 0-486-43945-3

BASIC ALGEBRA I: Second Edition, Nathan Jacobson. A classic text and standard reference for a generation, this volume covers all undergraduate algebra topics, including groups, rings, modules, Galois theory, polynomials, linear algebra, and associative algebra. 1985 edition. 528pp. 6 1/8 x 9 1/4. 0-486-47189-6

BASIC ALGEBRA II: Second Edition, Nathan Jacobson. This classic text and standard reference comprises all subjects of a first-year graduate-level course, including in-depth coverage of groups and polynomials and extensive use of categories and functors. 1989 edition. 704pp. 6 1/8 x 9 1/4. 0-486-47187-X

CALCULUS: An Intuitive and Physical Approach (Second Edition), Morris Kline. Application-oriented introduction relates the subject as closely as possible to science with explorations of the derivative; differentiation and integration of the powers of x; theorems on differentiation, antidifferentiation; the chain rule; trigonometric functions; more. Examples. 1967 edition. 960pp. 6 1/2 x 9 1/4. 0-486-40453-6

ABSTRACT ALGEBRA AND SOLUTION BY RADICALS, John E. Maxfield and Margaret W. Maxfield. Accessible advanced undergraduate-level text starts with groups, rings, fields, and polynomials and advances to Galois theory, radicals and roots of unity, and solution by radicals. Numerous examples, illustrations, exercises, appendixes. 1971 edition. 224pp. 6 1/8 x 9 1/4. 0-486-47723-1

AN INTRODUCTION TO THE THEORY OF LINEAR SPACES, Georgi E. Shilov. Translated by Richard A. Silverman. Introductory treatment offers a clear exposition of algebra, geometry, and analysis as parts of an integrated whole rather than separate subjects. Numerous examples illustrate many different fields, and problems include hints or answers. 1961 edition. 320pp. 5 3/8 x 8 1/2. 0-486-63070-6

LINEAR ALGEBRA, Georgi E. Shilov. Covers determinants, linear spaces, systems of linear equations, linear functions of a vector argument, coordinate transformations, the canonical form of the matrix of a linear operator, bilinear and quadratic forms, and more. 387pp. 5 3/8 x 8 1/2. 0-486-63518-X

Mathematics–Geometry and Topology

PROBLEMS AND SOLUTIONS IN EUCLIDEAN GEOMETRY, M. N. Aref and William Wernick. Based on classical principles, this book is intended for a second course in Euclidean geometry and can be used as a refresher. More than 200 problems include hints and solutions. 1968 edition. 272pp. 5 3/8 x 8 1/2. 0-486-47720-7

TOPOLOGY OF 3-MANIFOLDS AND RELATED TOPICS, Edited by M. K. Fort, Jr. With a New Introduction by Daniel Silver. Summaries and full reports from a 1961 conference discuss decompositions and subsets of 3-space; n-manifolds; knot theory; the Poincaré conjecture; and periodic maps and isotopies. Familiarity with algebraic topology required. 1962 edition. 272pp. 6 1/8 x 9 1/4. 0-486-47753-3

POINT SET TOPOLOGY, Steven A. Gaal. Suitable for a complete course in topology, this text also functions as a self-contained treatment for independent study. Additional enrichment materials make it equally valuable as a reference. 1964 edition. 336pp. 5 3/8 x 8 1/2. 0-486-47222-1

INVITATION TO GEOMETRY, Z. A. Melzak. Intended for students of many different backgrounds with only a modest knowledge of mathematics, this text features self-contained chapters that can be adapted to several types of geometry courses. 1983 edition. 240pp. 5 3/8 x 8 1/2. 0-486-46626-4

TOPOLOGY AND GEOMETRY FOR PHYSICISTS, Charles Nash and Siddhartha Sen. Written by physicists for physics students, this text assumes no detailed background in topology or geometry. Topics include differential forms, homotopy, homology, cohomology, fiber bundles, connection and covariant derivatives, and Morse theory. 1983 edition. 320pp. 5 3/8 x 8 1/2. 0-486-47852-1

BEYOND GEOMETRY: Classic Papers from Riemann to Einstein, Edited with an Introduction and Notes by Peter Pesic. This is the only English-language collection of these 8 accessible essays. They trace seminal ideas about the foundations of geometry that led to Einstein's general theory of relativity. 224pp. 6 1/8 x 9 1/4. 0-486-45350-2

GEOMETRY FROM EUCLID TO KNOTS, Saul Stahl. This text provides a historical perspective on plane geometry and covers non-neutral Euclidean geometry, circles and regular polygons, projective geometry, symmetries, inversions, informal topology, and more. Includes 1,000 practice problems. Solutions available. 2003 edition. 480pp. 6 1/8 x 9 1/4. 0-486-47459-3

TOPOLOGICAL VECTOR SPACES, DISTRIBUTIONS AND KERNELS, François Trèves. Extending beyond the boundaries of Hilbert and Banach space theory, this text focuses on key aspects of functional analysis, particularly in regard to solving partial differential equations. 1967 edition. 592pp. 5 3/8 x 8 1/2. 0-486-45352-9

INTRODUCTION TO PROJECTIVE GEOMETRY, C. R. Wylie, Jr. This introductory volume offers strong reinforcement for its teachings, with detailed examples and numerous theorems, proofs, and exercises, plus complete answers to all odd-numbered end-of-chapter problems. 1970 edition. 576pp. 6 1/8 x 9 1/4. 0-486-46895-X

FOUNDATIONS OF GEOMETRY, C. R. Wylie, Jr. Geared toward students preparing to teach high school mathematics, this text explores the principles of Euclidean and non-Euclidean geometry and covers both generalities and specifics of the axiomatic method. 1964 edition. 352pp. 6 x 9. 0-486-47214-0

Mathematics–History

THE WORKS OF ARCHIMEDES, Archimedes. Translated by Sir Thomas Heath. Complete works of ancient geometer feature such topics as the famous problems of the ratio of the areas of a cylinder and an inscribed sphere; the properties of conoids, spheroids, and spirals; more. 326pp. 5 3/8 x 8 1/2. 0-486-42084-1

THE HISTORICAL ROOTS OF ELEMENTARY MATHEMATICS, Lucas N. H. Bunt, Phillip S. Jones, and Jack D. Bedient. Exciting, hands-on approach to understanding fundamental underpinnings of modern arithmetic, algebra, geometry and number systems examines their origins in early Egyptian, Babylonian, and Greek sources. 336pp. 5 3/8 x 8 1/2. 0-486-25563-8

THE THIRTEEN BOOKS OF EUCLID'S ELEMENTS, Euclid. Contains complete English text of all 13 books of the Elements plus critical apparatus analyzing each definition, postulate, and proposition in great detail. Covers textual and linguistic matters; mathematical analyses of Euclid's ideas; classical, medieval, Renaissance and modern commentators; refutations, supports, extrapolations, reinterpretations and historical notes. 995 figures. Total of 1,425pp. All books 5 3/8 x 8 1/2.
Vol. I: 443pp. 0-486-60088-2
Vol. II: 464pp. 0-486-60089-0
Vol. III: 546pp. 0-486-60090-4

A HISTORY OF GREEK MATHEMATICS, Sir Thomas Heath. This authoritative two-volume set that covers the essentials of mathematics and features every landmark innovation and every important figure, including Euclid, Apollonius, and others. 5 3/8 x 8 1/2. Vol. I: 461pp. 0-486-24073-8
Vol. II: 597pp. 0-486-24074-6

A MANUAL OF GREEK MATHEMATICS, Sir Thomas L. Heath. This concise but thorough history encompasses the enduring contributions of the ancient Greek mathematicians whose works form the basis of most modern mathematics. Discusses Pythagorean arithmetic, Plato, Euclid, more. 1931 edition. 576pp. 5 3/8 x 8 1/2.
0-486-43231-9

CHINESE MATHEMATICS IN THE THIRTEENTH CENTURY, Ulrich Libbrecht. An exploration of the 13th-century mathematician Ch'in, this fascinating book combines what is known of the mathematician's life with a history of his only extant work, the Shu-shu chiu-chang. 1973 edition. 592pp. 5 3/8 x 8 1/2.
0-486-44619-0

PHILOSOPHY OF MATHEMATICS AND DEDUCTIVE STRUCTURE IN EUCLID'S ELEMENTS, Ian Mueller. This text provides an understanding of the classical Greek conception of mathematics as expressed in Euclid's Elements. It focuses on philosophical, foundational, and logical questions and features helpful appendixes. 400pp. 6 1/2 x 9 1/4. 0-486-45300-6

BEYOND GEOMETRY: Classic Papers from Riemann to Einstein, Edited with an Introduction and Notes by Peter Pesic. This is the only English-language collection of these 8 accessible essays. They trace seminal ideas about the foundations of geometry that led to Einstein's general theory of relativity. 224pp. 6 1/8 x 9 1/4. 0-486-45350-2

HISTORY OF MATHEMATICS, David E. Smith. Two-volume history – from Egyptian papyri and medieval maps to modern graphs and diagrams. Non-technical chronological survey with thousands of biographical notes, critical evaluations, and contemporary opinions on over 1,100 mathematicians. 5 3/8 x 8 1/2.
Vol. I: 618pp. 0-486-20429-4
Vol. II: 736pp. 0-486-20430-8